Springer Theses

Recognizing Outstanding Ph.D. Research

Aims and Scope

The series "Springer Theses" brings together a selection of the very best Ph.D. theses from around the world and across the physical sciences. Nominated and endorsed by two recognized specialists, each published volume has been selected for its scientific excellence and the high impact of its contents for the pertinent field of research. For greater accessibility to non-specialists, the published versions include an extended introduction, as well as a foreword by the student's supervisor explaining the special relevance of the work for the field. As a whole, the series will provide a valuable resource both for newcomers to the research fields described, and for other scientists seeking detailed background information on special questions. Finally, it provides an accredited documentation of the valuable contributions made by today's younger generation of scientists.

Theses are accepted into the series by invited nomination only and must fulfill all of the following criteria

- They must be written in good English.
- The topic should fall within the confines of Chemistry, Physics, Earth Sciences, Engineering and related interdisciplinary fields such as Materials, Nanoscience, Chemical Engineering, Complex Systems and Biophysics.
- The work reported in the thesis must represent a significant scientific advance.
- If the thesis includes previously published material, permission to reproduce this must be gained from the respective copyright holder.
- They must have been examined and passed during the 12 months prior to nomination.
- Each thesis should include a foreword by the supervisor outlining the significance of its content.
- The theses should have a clearly defined structure including an introduction accessible to scientists not expert in that particular field.

More information about this series at http://www.springer.com/series/8790

Olga Isupova

Machine Learning Methods for Behaviour Analysis and Anomaly Detection in Video

Doctoral Thesis accepted by
the University of Sheffield, Sheffield, UK

Springer

Author
Dr. Olga Isupova
Department of Engineering Science
University of Oxford
Oxford
UK

Supervisor
Prof. Lyudmila Mihaylova
University of Sheffield
Sheffield
UK

ISSN 2190-5053 ISSN 2190-5061 (electronic)
Springer Theses
ISBN 978-3-030-09250-4 ISBN 978-3-319-75508-3 (eBook)
https://doi.org/10.1007/978-3-319-75508-3

Printed on acid-free paper

This Springer imprint is published by the registered company Springer International Publishing AG part of Springer Nature
The registered company address is: Gewerbestrasse 11, 6330 Cham, Switzerland

Supervisor's Foreword

Autonomy implies safety, robustness and quick decision-making. With increasing amounts of data, it becomes almost impossible for a human to analyse and process these vast data streams. Volume, velocity, veracity and variety are some of the challenges that autonomy brings. Despite the significant progress made in this area, there is still a big gap between the required level of autonomy and what the technology currently provides.

Autonomous change detection and anomaly detection are needed in many areas, from smartphones to drones, autonomous cars to cybersecurity, and video analytical systems. In this context, the Ph.D. thesis of Olga Isupova makes important theoretical and methodological contributions. It presents machine learning approaches able to detect changes in video streams and to replace the human involvement.

The thesis develops Bayesian nonparametric inference methods for anomaly detection and behaviour analysis and represents a step change compared with existing approaches from this actively investigated area in several aspects. The methods involve elements of learning which consist of normal behaviour and are able to detect anything contradicting this normal behaviour. The thesis makes significant progress in processing high-dimensional video data and provides a topic modelling approach for detecting any kind of anomalies, without requiring these to be specified in advance.

The thesis proposes hierarchical Dirichlet process state models which assume that data consists of an unknown number, with potentially infinite series of switching activities and behaviours. When newly upcoming data contradicts learnt series of behaviours, an anomaly is detected.

The second significant contribution of the thesis is in the developed Gaussian process method for change detection and in corroborating the theoretical underpinnings. Both methods have been applied to anomaly detection and localisation in video streams, and their efficiency for autonomous online processing has been demonstrated. The proposed methods are general and can be applied to various applications where data does not have a clear structure nor easily predictable representations.

In addition, the thesis is accompanied with open-source toolboxes which can be used by researchers and engineers.

I am extremely grateful to *Springer Theses Series* for providing this opportunity so that the research results of Olga Isupova can reach a wide audience.

Sheffield, UK
November 2017

Prof. Lyudmila Mihaylova

Abstract

Behaviour analysis and anomaly detection are key components of intelligent vision systems. Anomaly detection can be considered from two perspectives: abnormal events can be defined as those that violate typical activities or as a sudden change in behaviour. Topic modelling and change point detection methodologies, respectively, are employed to achieve these objectives.

The thesis starts with development of novel learning algorithms for a dynamic topic model. Topics extracted by the learning algorithms represent typical activities happening within an observed scene. These typical activities are used for likelihood computation. The likelihood serves as a normality measure in anomaly detection decision-making. A novel anomaly localisation procedure is proposed.

In the considered dynamic topic model, a number of topics, i.e. typical activities, should be specified in advance. A novel dynamic nonparametric hierarchical Dirichlet process topic model is then developed where the number of topics is determined from data. Conventional posterior inference algorithms require processing of the whole data through several passes. It is computationally intractable for massive or sequential data. Therefore, batch and online inference algorithms for the proposed model are developed. A novel normality measure is derived for decision-making in anomaly detection.

The latter part of the thesis considers behaviour analysis and anomaly detection within the change point detection methodology. A novel general framework for change point detection is introduced. Gaussian process time series data is considered, and a change is defined as an alteration in hyperparameters of the Gaussian process prior. The problem is formulated in the context of statistical hypothesis testing, and several tests suitable for both offline and online data processing and multiple change point detection are proposed. Theoretical properties of the proposed tests are derived based on the distribution of the test statistics.

Acknowledgements

I am extremely grateful to my supervisor Prof. Lyudmila Mihaylova for the encouragement of my independent research. She also gave me significant freedom in choosing a research direction and methodology. She has always been very enthusiastic if I come up with an idea to try a new approach or to study a new topic. I respect the width of her research interests. It has been especially helpful that Mila has never given me an opportunity to lose concentration, making me constantly do my best. I would also like to thank Mila for her support in my non-academic life, for her useful advice and for her care in situations beyond my Ph.D. journey.

I would like to acknowledge Prof. Fredrik Gustafsson who made the part of this thesis devoted to change point detection possible. I admire how elegantly he brought my heuristic ideas to a considerably new theoretical level. It was my honour and pleasure to work with Fredrik, and I got an invaluable experience from my visit. I would also like to thank Dr. Carsten Fritsche and the other members of the Automatic Control Division at the Linkoping University for organising my activities during this visit.

My colleagues at Mila's group—Allan de Freitas, Matthew Hawes, Naveed Salman and Ata ur-Rehman—have helped me throughout this journey. I appreciate our discussions and exchange of experience.

I would also like to thank my master's thesis supervisor, Prof. Konstantin Vorontsov, for inculcating my endless interest in topic modelling and demonstrating how fascinating a research work can be. His passion to research made me start this Ph.D.

I acknowledge the EC Seventh Framework Programme [FP7 2013–2017] TRAcking in compleX sensor systems (TRAX) Grant agreement no.: 607400 for funding my research.

I am not able to put into words how grateful I am to my husband, Danil Kuzin. Without him I would not be in this stage now. I would like to thank him for his research collaboration, invaluably useful and indispensable discussions, patience and support. He has always challenged me making me permanently improve and develop myself.

Contents

Acronyms

CUSUM	Cumulative Sum algorithm
EM	Expectation-Maximisation
GP	Gaussian Process
GP-BOCPD	Gaussian Process Bayesian Online Change Point Detection
GP-glLRT	Gaussian Process generalised log Likelihood Ratio Test
GP-lLRT	Gaussian Process log Likelihood Ratio Test
GP-OLCDT	Gaussian Process Online Likelihood-based Change point Detection Test
GS	Gibbs Sampling
HDP	Hierarchical Dirichlet Process
LDA	Latent Dirichlet Allocation
LRT	Likelihood Ratio Test
MAP	Maximum a Posteriori
MCTM	Markov Clustering Topic Model
MLE	Maximum Likelihood Estimates
PLSA	Probabilistic Latent Semantic Analysis
VB	Variational Bayes

Notation

Joint Notations for Topic Modelling

$\mathcal{J}$	A set of visual documents
$\mathcal{V}$	A vocabulary of visual words
$\mathcal{K}$	A set of topics
N_j	A length of visual document j
J	The number of visual documents, and it is assumed that $\mathcal{J} = \{1, \ldots, J\}$
V	A size of word vocabulary
j	A visual document, $j \in \mathcal{J}$
w	A visual word, $w \in \mathcal{V}$
k	A topic, $k \in \mathcal{K}$
w_{ji}	A visual word at position i in visual document j
$\mathbf{w}_j$	A set of visual words in visual document j, $\mathbf{w}_j = \{w_{ji}\}_{i=1}^{N_j}$
$\mathbf{w}_{j':j''}$	A set of visual words in the sequence of visual documents starting from the j'th document till the document j'', $\mathbf{w}_{j':j''} = \{\mathbf{w}_j\}_{j=j'}^{j''}$
z_{ji}	A topic assigned to position i in visual document j
$\mathbf{z}_j$	A set of topic assignments in visual document j, $\mathbf{z}_j = \{z_{ji}\}_{i=1}^{N_j}$
$\mathbf{z}_{j':j''}$	A set of topic assignments in the sequence of visual documents starting from the j'th document till the document j'', $\mathbf{z}_{j':j''} = \{\mathbf{x}_j\}_{j=j'}^{j''}$
ϕ_{wk}	A probability of word w in topic k, $\phi_{wk} = p(w\|k)$
$\boldsymbol{\phi}_k$	A distribution over words in topic k, $\boldsymbol{\phi}_k = \{\boldsymbol{\phi}_{wk}\}_{w \in \mathcal{V}}$
$\boldsymbol{\Phi}$	A matrix of distributions over words for all the topics, $\boldsymbol{\Phi} = \{\boldsymbol{\phi}_k\}_{k \in \mathcal{K}}$
θ_{kj}	A probability of topic k in document j, $\theta_{kj} = p(k\|j)$
$\boldsymbol{\theta}_j$	A distribution over topics in document j, $\boldsymbol{\theta}_j = \{\theta_{kj}\}_{k \in \mathcal{K}}$
$\boldsymbol{\Theta}$	A matrix of distribution over topics for all documents, $\boldsymbol{\Theta} = \{\boldsymbol{\theta}_j\}_{j=\in \mathcal{J}}$
n_{wj}	A counter for the number of times word w appears in document j
l_{wk}	A counter for the number of times word w is associated with topic k

$\boldsymbol{\eta}$	A hyperparameter of a prior Dirichlet distribution for distributions $\boldsymbol{\phi}_k$
$\boldsymbol{\alpha}$	A hyperparameter of a prior Dirichlet distribution for distributions $\boldsymbol{\theta}_j$
$\mathcal{A}(\cdot)$	A normality measure

Notations Specific for the MCTM

$\mathcal{B}$	A set of behaviours
b	A behaviour, $b \in \mathcal{B}$
b_j	A behaviour assigned to visual document j
$\mathbf{b}_{j':j''}$	A set of behaviour assignments in the sequence of visual documents starting from the j'th document till the document j'', $\mathbf{b}_{j':j''} = \{b_j\}_{j=j'}^{j''}$
θ_{kb}	A probability of topic k in behaviour b, $\theta_{kb} = p(k\|b)$[1]
$\boldsymbol{\theta}_b$	A distribution over topics in behaviour b, $\boldsymbol{\theta}_b = \{\theta_{kb}\}_{k \in \mathcal{K}}$
$\boldsymbol{\Theta}$	A matrix of distribution over topics for all behaviours, $\boldsymbol{\Theta} = \{\boldsymbol{\theta_b}\}_{b \in \mathcal{B}}$
$\xi_{b'b}$	A transitional probability to switch from behaviour b to behaviour b', $\xi_{b'b} = p(b'\|b)$
$\boldsymbol{\xi}_b$	A transitional distribution to switch from behaviour b, $\boldsymbol{\xi}_b = \{\xi_{b'b}\}_{b' \in \mathcal{B}}$
$\boldsymbol{\Xi}$	A transition probability matrix for a Markov chain of behaviours, $\boldsymbol{\Xi} = \{\boldsymbol{\xi}_b\}_{b \in \mathcal{B}}$
ω_b	A probability of behaviour b to be associated with the first document, $\omega_b = p(b)$
$\boldsymbol{\omega}$	A distribution over behaviours to be associated with the first document, $\boldsymbol{\omega} = \{\omega_b\}_{b \in \mathcal{B}}$
n_{kb}	A counter for the number of times topic k is associated with behaviour b
$n_{b'b}$	A counter for the number of times behaviour b is followed by behaviour b'
n_b	A counter for the number of times behaviour b is associated with the first document
$\boldsymbol{\alpha}$	A hyperparameter of a prior Dirichlet distribution for distributions $\boldsymbol{\theta}_b$
$\boldsymbol{\upsilon}$	A hyperparameter of a prior Dirichlet distribution for distributions $\boldsymbol{\xi}_b$
$\boldsymbol{\varkappa}$	A hyperparameter of a prior Dirichlet distribution for the distribution $\boldsymbol{\omega}$
$\boldsymbol{\Omega}$	A set of parameters of the MCTM, $\boldsymbol{\Omega} = \{\boldsymbol{\Phi}, \boldsymbol{\Theta}, \omega, \boldsymbol{\Xi}\}$
$\mathcal{Q}(\cdot)$	The expected logarithm of the full likelihood of observed and hidden variables that is used in the EM-algorithm

Notations Specific for the Dynamic HDP

G_0	A top-level Dirichlet process
G_j	A document-level Dirichlet process
H	A base measure for the top-level Dirichlet process
n_{jt}	A counter for the number of words assigned to table t in document j

[1] In the MCTM, a topic mixture is associated with a behaviour rather than with a document as in conventional topic models.

m_{jk}	A counter for the number of tables that have topic k in document j
t	A table in a document-level Chinese restaurant, i.e. a document
t_{ji}	A table assigned to the ith word in document j
$\mathbf{t}_{j':j''}$	A set of table assignments in the sequence of visual documents starting from the j'th document till the document j'', $\mathbf{t}_{j':j''} = \{t_{ji}\}_{j=j':j'',i=1:N_j}$
k_{jt}	A topic assigned to table t in document j
$\mathbf{k}_{j':j''}$	A set of topic assignments in the sequence of visual documents starting from the j'th document till the document j'', $\mathbf{k}_{j':j''} = \{k_{jt}\}_{j=j':j'',t=1:m_{j\cdot}}$
α	A concentration parameter of a document-level Dirichlet process G_j
γ	A concentration parameter of a top-level Dirichlet process G_0
λ	A weighted parameter for topics used in an entire data set

Notations for Change Point Detection

N	The number of points in a training data set
τ	A time index
$\boldsymbol{\tau}_{i':i''}$	A vector of time indices, $\boldsymbol{\tau}_{i':i''} = \{\tau_i\}_{i=i'}^{i''}$
y_τ	An observation at time τ
$\mathbf{y}_{i':i''}$	Observations at the given time indices, $\mathbf{y}_{i':i''} = \{y_{\tau_i}\}_{i=i'}^{i''}$
$f(\cdot)$	A time series function
$\mathbf{f}_{i':i''}$	Function values for the given time indices, $\mathbf{f}_{i':i''} = f(\boldsymbol{\tau}_{i':i''}) = \{f(\tau_i)\}_{i=i'}^{i''}$
$m(\cdot)$	A mean function of a Gaussian process
$\boldsymbol{\mu}$	Realisations of the Gaussian process mean function at the input points, $\boldsymbol{\mu} = \{\mu_i\}_{i=1}^N = \{m(\tau_i)\}_{i=1}^N$
$k(\cdot,\cdot)$	A covariance function of a Gaussian process
$\mathbf{K}$	Realisations of the Gaussian process covariance function at the input points, $\mathbf{K} = \{\mathbf{K}_{i,j}\}_{i,j=1}^N = \{k(\tau_i,\tau_j)\}_{i,j=1}^N$
ε_τ	Additive noise of Gaussian process observations
σ^2	A variance of the additive noise
ϑ	A hyperparameter vector of a Gaussian process
$\mathbf{I}$	The identity matrix
$\mathcal{H}_0$	A null hypothesis
$\mathcal{H}_1$	An alternative
τ^*	Change time

Functions and Distributions

$\psi(\cdot)$	The digamma function
$\Gamma(\cdot)$	The gamma function
$Cat(\cdot)$	A categorical distribution
$Dir(\cdot)$	A Dirichlet distribution
$\mathcal{DP}(\cdot,\cdot)$	A Dirichlet process
$\mathcal{N}(\cdot,\cdot)$	A Gaussian distribution

$\mathcal{GP}(\cdot,\cdot)$	A Gaussian process
χ^2_n	A chi-squared distribution with n degrees of freedom
$\chi'^2_n(\cdot)$	A non-central chi-squared distribution with n degrees of freedom

List of Figures

List of Tables

Chapter 1
Introduction

Intelligent video systems and analytics represent an active research field combining methods from computer vision, machine learning, data mining, signal processing and other areas for mining meaningful information from raw video data. The availability of cheap sensors and need for solving intelligent tasks facilitate the growth of interest in this area. Vast amount of data collected by different devices require automatic systems for analysis. These systems should be able to make decisions without human interruption or with minimal assistance from a human operator. Video analytics systems should understand and interpret a scene, detect motion, classify and track objects, explore typical behaviours and detect abnormal events [1].

The application area of such systems is huge: preventing crimes in public spaces such as airports, railway stations, or schools; counting objects at stadiums or shopping malls; detection of breaks or leaks; smart homes for elderly people maintenance with fall detection functionality and others.

Behaviour analysis and anomaly detection are essential parts of intelligent video systems [2, 3]. The objectives of anomaly detection are to detect and inform about any unusual, suspicious and abnormal events happening within the observed scene. These may be pedestrians crossing a road in a wrong place, cars running on the red light, abandoned objects, a person fall, a pipe leak and others. Decisions made by a system should be interpretable by a human therefore the system should also provide information about typical behaviours to confirm its decisions.

This thesis develops machine learning methods for automatic behaviour analysis and anomaly detections in video. The methods allow to extract semantic patterns from data. These patterns can be interpreted as behaviours and they are used as a basis for decision making in anomaly detection.

O. Isupova, *Machine Learning Methods for Behaviour Analysis and Anomaly Detection in Video*, Springer Theses, https://doi.org/10.1007/978-3-319-75508-3_1

1.1 Abnormal Behaviour Detection

The earliest video systems required permanent monitoring due to impossibility of data recording. Despite the fact of significant improvements of technologies human operators still perform anomaly detection. Human labour is expensive and exhausting due to the fact that most of the time nothing important happens and operators lose concentration. Moreover, human monitoring is unable to cover a significantly growing amount of data. This leads to necessity of automated anomaly detection systems. However, even a formal definition of abnormality is challenging, as it should include such informal concepts as "interesting", "unusual", "suspicious".

These systems should satisfy the number of requirements:

- **Autonomous decision making**. In real world applications it is impossible to predict a priori all kinds of abnormalities that can happen. The system should then be self-learning and be able to work without data labels.
- **Interpretability**. The decisions based on anomaly detections made by the system can be crucial and affect people. Moreover, actions required by anomaly alerts can be expensive to take. Therefore, humans should trust the system and understand why the system makes a specific decision. Black-box algorithms are hence inappropriate for a reliable anomaly detection system. Behaviour analysis can then be employed jointly with anomaly detection to provide required explanation of system decisions.
- **Real time execution**. The system should detect an anomaly immediately to warn human security staff. In some applications, such as crime investigation, offline systems are sufficient to analyse past data. However, online systems are more critical, they can replace current human operators. Therefore, an algorithm for anomaly detection should be efficient.

This thesis presents methods to address these criteria. We employ the probabilistic framework to model and interpret behaviours within the scene and to detect anomalies. Two approaches are considered: topic modeling to extract typical local motion patterns and to detect activities that violate these patterns (e.g., a car running on the red traffic light) and change point detection to identify global anomalies that break a normal motion stream (e.g., panic in a crowd).

1.1.1 Topic Modeling

Topic modeling [4] methods extract a latent structure in observed data allowing to represent it in a low-dimensional space of so-called *topics*. Topics form sets of features that statistically often appear together. Therefore, topics can be used to express normal patterns. In video applications a topic can represent a typical local motion pattern or *activity* (Fig. 1.1). Note that topics would represent only normal activities and do not capture abnormal ones even though they are present in data. In this context abnormal activities are understood as those that happen rarely.

(a) (b)

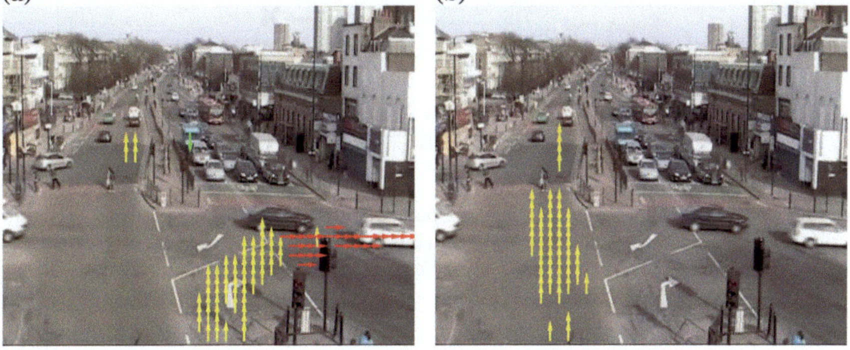

Fig. 1.1 Samples of typical activities discovered by a topic model on busy road junction data

Likelihood of new observations based on extracted topics is a reasonable normality measure within the topic modeling approach. A likelihood value is low when an observation does not fit to the learnt topics, i.e., normal activities. Therefore, a low likelihood value would indicate that something atypical or abnormal happens.

Topic modeling addresses all the requirements for anomaly detection systems formulated above. As topic modeling is fully unsupervised it does not require any labels in data, it does not even assume that all observations in a training dataset are normal. Topics provide descriptions of normal activities that help in explanation of anomaly detection decisions. Recent advances in topic modeling allow to apply topic models on big and streaming data efficiently, e.g., [5, 6].

The first part of the thesis is devoted to development of methods for behaviour analysis and anomaly detection within the topic modeling approach.

1.1.2 Change Point Detection

Change point detection represents an alternative approach for behaviour analysis and anomaly detection in video. Topic modeling methods extract typical behaviours and detect abnormal events that cannot be explained by these behaviours. Change point detection can be applied for discovering sudden changes in global behaviours. For example, it is useful for crowd panic detection in public places or for behaviour segmentation in smart homes. In contrast to topic modeling change point detection can be employed in unknown situations.

Change point detection [7] methods find changes in a probability distribution of stochastic processes or time series. In the context of behaviour analysis different probability distributions represent different behaviours (Fig. 1.2). If at the beginning of observation, a normal behaviour is expected, then a change in the distribution may mean an anomaly.

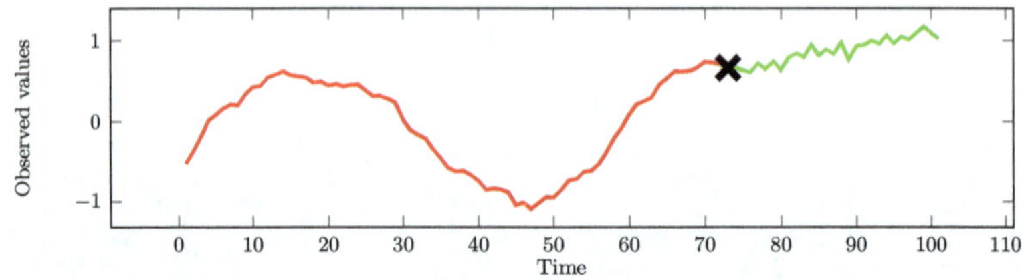

Fig. 1.2 Sample of a time series with a change point that separates different behaviours. The time series represents coordinates of a dancing bee. In this figure two dance phases are given: right turn (red) and waggle (green). A cross indicates the change point. Details of the dance bee data are provided in Chap. 5

The requirements for anomaly detection systems are also fulfilled within the change point detection approach. Change point detection methods can work in an unsupervised manner where parameters of underlying distributions are estimated from observed data. Interpretability is ensured by the correspondence between behaviours and probability distributions of observed data. Most of the methods for change point detection are online and designed for quickest detection.

The developed methods for behaviour analysis and anomaly detection within the change point detection approach are presented in the latter part of the thesis.

1.2 Key Contributions and Outline

Below is a brief overview of the content and main contributions presented in the thesis chapters.

Chapter 2. This chapter reviews the concepts and algorithms that serve as a basis for the methods proposed in this thesis. Knowledge mining from video starts from an analysis of video data and extraction of informative features from it, therefore the review starts from outlining the methods for video processing. The chapter then covers the current state of the art in the area of anomaly detection in video. This thesis considers two approaches for anomaly detection via topic modeling and change point detection. An overview of both areas is presented at the end of the chapter.

Chapter 3. A dynamic topic model for behaviour analysis and anomaly detection in video is considered in this chapter. The focus of this chapter is on development and comparison of different learning algorithms for the topic model. Predictive likelihood of newly observed data is used as a normality measure for decision making. A novel procedure to localise a detected anomaly is proposed in the chapter. The key contributions of this chapter consist of:

- New learning algorithms for the Markov clustering topic model are developed.
- An anomaly localisation procedure that follows concepts of probabilistic topic modeling is designed.
- Likelihood expressions as a normality measure of newly observed data are derived.
- Comprehensive analysis of the algorithms over real video sequences is introduced. Experiments show that the proposed methods provide more accurate results than the previously developed learning method in terms of anomaly detection performance. The experiments also confirm effectiveness of the proposed anomaly localisation procedure.

Chapter 4. A novel dynamic nonparametric topic model is proposed in this chapter. In contrast to the model, analysed in the previous chapter, the number of model parameters is not fixed in advance and, in practice, it is determined from data. The model is general and can be applied on any kind of data, where one is interested in extracting typical patterns from dynamic data. The dynamics are assumed such that mixtures of typical patterns at successive data points are similar. In this chapter the model is considered in the context of behaviour analysis and anomaly detection in video. The key contributions of this chapter are as follows:

- A novel dynamic nonparametric topic model is designed.
- An inference scheme that is combination of batch and online data processing is developed.
- A normality measure for anomaly detection decision making is proposed.
- The introduced method is evaluated on both synthetic and real video data. The results show that consideration of dynamics in a model significantly improves anomaly detection performance.

Chapter 5. This chapter considers the problem of behaviour analysis and anomaly detection in video from a different perspective in comparison to the previous chapters. A video is presented as a time series and the change point detection methodology is employed for it. A novel general framework for change point detection is introduced in this chapter. The Gaussian process time series model is considered. The problem is formulated in the context of statistical hypothesis testing. Statistical tests proposed in this chapter can be applied for both offline and online data processing to detect single and multiple change points. The key contributions of this chapter can be summarised as:

- The change point detection problem is formulated within the statistical hypothesis testing approach, where a change is defined as an alteration in hyperparameters of a Gaussian process prior.
- A general framework for change point detection is proposed.
- Statistical tests for change point detection are defined and developed.
- Theoretical properties of the introduced statistical tests are derived.
- A change point detection method for online data processing able to detect multiple change points is designed.

- The proposed methods are thoroughly evaluated in terms of the typical change point detection measures on synthetic and real video data. Results achieved by the developed methods demonstrate a tradeoff between false alarm and missed detection rates.

Chapter 6. The chapter presents an overview of the main results of the thesis and directions for future work based on the research provided in the thesis.

1.3 Disseminated Results

The results presented in this thesis are disseminated in the following papers:
Journal papers

- O. Isupova, D. Kuzin, L. Mihaylova."Learning Methods for Dynamic Topic Modeling in Automated Behaviour Analysis", in *IEEE Transactions on Neural Networks and Learning Systems*, vol. PP, no. 99, pp. 1–14
- O. Isupova, D. Kuzin, F. Gustafsson, L. Mihaylova. "Change Point Detection with Gaussian Processes", in *IEEE Transactions on Pattern Analysis and Machine Intelligence*, under review, 2017

Peer-reviewed conferences

- O. Isupova, D. Kuzin, L. Mihaylova. "Dynamic Hierarchical Dirichlet Process for Abnormal Behaviour Detection in Video", in *Proceedings of the 19th International Conference on Information Fusion*, 5–8 July 2016, Heidelberg, Germany, pp. 750–757
- O. Isupova, L. Mihaylova, D. Kuzin, G. Markarian, F. Septier. "An Expectation Maximisation Algorithm for Behaviour Analysis in Video", in *Proceedings of 18th International Conference on Information Fusion*, 6–9 July 2015, Washington, USA, pp. 126–133
- O. Isupova, D. Kuzin, L. Mihaylova. "Abnormal Behaviour Detection in Video Using Topic Modeling", in *Proceedings of University of Sheffield Engineering Symposium*, 24 June 2015, Sheffield, UK

Workshops

- O. Isupova, D. Kuzin, L. Mihaylova. "Anomaly Detection in Video with Bayesian Nonparametrics", in *ICML 2016 Anomaly detection Workshop*, 24 June 2016, New York, NY, USA

The following papers have been published covering the related work:

Peer-reviewed conferences

- D. Kuzin, O. Isupova, L.Mihaylova. "Structured Sparse Modelling with Hierarchical GP" in *Proceedings of the Signal Processing with Adaptive Sparse Structured Representations (SPARS) Workshop, 5–8 June 2017, Lisbon, Portugal*

- Z.Li, O. Isupova, L. Mihaylova, L. Rossi. "Autonomous Flame Detection in Video Based on Saliency Analysis and Optical Flow", in *Proceedings of the IEEE International Conference on Multisensor Fusion and Integration for Intelligent Systems*, 19–21 September 2016, Baden-Baden, Germany, pp. 218–223
- D. Kuzin, O. Isupova, L. Mihaylova. "Compressive Sensing Approaches for Autonomous Object Detection in Video Sequences", in *Proceedings of the 10th Workshop Sensor Data Fusion: Trends, Solutions, and Applications*, 6–8 October 2015, Bonn, Germany, pp. 1–6

References

1. H. Liu, S. Chen, N. Kubota, Intelligent video systems and analytics: a survey. IEEE Trans. Industr. Inf. **9**(3), 1222–1233 (2013)
2. T. Li, H. Chang, M. Wang, B. Ni, R. Hong, S. Yan, Crowded scene analysis: a survey. IEEE Trans. Circuits Syst. Video Technol. **25**(3), 367–386 (2015)
3. P.V.K. Borges, N. Conci, A. Cavallaro, Video-based human behavior understanding: a survey. IEEE Trans. Circuits Syst. Video Technol. **23**(11), 1993–2008 (2013)
4. D.M. Blei, Probabilistic topic models. Commun. ACM **55**(4), 77–84 (2012)
5. M.D. Hoffman, D.M. Blei, C. Wang, J.W. Paisley, Stochastic variational inference. J. Mach. Learn. Res. **14**(1), 1303–1347 (2013)
6. P. Smyth, M. Welling, A.U. Asuncion, Asynchronous distributed learning of topic models, in *Advances in Neural Information Processing Systems (NIPS 2009)*, pp. 81–88, 2009
7. M. Basseville, I.V. Nikiforov, *Detection of Abrupt Changes—Theory and Application* (Prentice Hall Inc., 1993)

Chapter 2
Background

In this chapter a review of related works is provided. Presented concepts and methods are used as a basis for the main contributions given in the later chapters. Video analysis starts from processing raw data. The objective is to develop algorithms and tools for autonomous systems that can extract knowledge from data. Section 2.1 gives a bird's sight view on the algorithms for video processing. Section 2.2 provides a survey of the current state of the art in the area of anomaly detection in video, covering each step of the data analysis. Section 2.3 delivers an introduction to the area of topic modeling that is used in Chaps. 3 and 4. Section 2.4 overviews the area of change point detection, which is considered in Chap. 5.

2.1 Outline of Video Processing Methods

First step in video processing is to determine objects of interest within the scene and estimate their motion. The following sections present a brief overview of the methods used for object detection and tracking in video.

2.1.1 Object Detection

There are a number of approaches for object detection depending on applications. The goal can be to detect any moving objects or detect some specially defined objects such as faces or vehicles.

O. Isupova, *Machine Learning Methods for Behaviour Analysis and Anomaly Detection in Video*, Springer Theses, https://doi.org/10.1007/978-3-319-75508-3_2

2.1.1.1 Optical Flow

The analysis of motion can significantly help in video processing and mining semantic information from it. Therefore, motion detection is an important step in video analytics. An optical flow is a vector field of apparent pixel motion between frames. Optical flow estimation is one of the basic methods for motion detection.

The problem of finding an optical flow between two frames can be formulated as follows:

Input:
I_1, I_2 — two frames of one scene;
Output:
$\{\mathbf{u}, \mathbf{v}\} = \{u_{\check{i},\check{j}}, v_{\check{i},\check{j}}\}_{\check{i}=1,\check{j}=1}^{\check{N},\check{M}}$ — the vector field of apparent pixel motion between the input frames, where $u_{\check{i},\check{j}}, v_{\check{i},\check{j}}$ are the horizontal and vertical displacements of the pixel with the coordinates $(\check{i}, \check{j})$ in the frame I_1, i.e., the new location of this pixel in the frame I_2 is $(\check{i} + u_{\check{i},\check{j}}, \check{j} + v_{\check{i},\check{j}})$; $\check{N}, \check{M}$ are the sizes of the frame

Methods for optical flow estimation can be classified as global or local. The global methods search for an optical flow $\{\mathbf{u}, \mathbf{v}\}$ for the entire image. The local methods seek a vector $\{u_{\check{i},\check{j}}, v_{\check{i},\check{j}}\}$ for each pixel.

2.1.1.2 Global Methods

The first global method for optical flow estimation was proposed in [1]. They use two basic assumptions:

1. **Colour consistency**. Pixels do not change colour while moving from one frame to another, i.e., $I_1(\check{i}, \check{j}) = I_2\left(\check{i} + u_{\check{i},\check{j}}, \check{j} + v_{\check{i},\check{j}}\right)$
2. **Spatial similarity**. Neighbouring pixels move similarly and have similar optical flow vectors, i.e., $\{u_{\check{i},\check{j}}, v_{\check{i},\check{j}}\} \approx \{u_{\tilde{i},\tilde{j}}, v_{\tilde{i},\tilde{j}}\}$, where $(\tilde{i}, \tilde{j})$ coordinates from the $(\check{i}, \check{j})$th pixel neighbourhood

Taking into account these assumptions a cost function for finding the global optical flow for the whole frame can be formulated as follows:

$$E(\mathbf{u}, \mathbf{v}) = \sum_{\check{i},\check{j}=1}^{\check{N},\check{M}} \Big[\rho_{\text{colour}}\left(I_1(\check{i}, \check{j}) - I_2(\check{i} + u_{\check{i},\check{j}}, \check{j} + v_{\check{i},\check{j}})\right) + \\ + \lambda_{\text{OF}}\Big(\rho_{\text{spatial}}(u_{\check{i},\check{j}} - u_{\check{i}+1,\check{j}}) + \rho_{\text{spatial}}(u_{\check{i},\check{j}} - u_{\check{i},\check{j}+1}) + \\ \rho_{\text{spatial}}(v_{\check{i},\check{j}} - v_{\check{i}+1,\check{j}}) + \rho_{\text{spatial}}(v_{\check{i},\check{j}} - v_{\check{i},\check{j}+1})\Big)\Big], \tag{2.1}$$

where ρ_{colour} and ρ_{spatial} are penalty functions, λ_{OF} is a regularisation parameter. An optimisation method is then applied to find the optimal optical flow.

Global methods of optical flow estimation differ in penalty functions, optimisation methods and additional heuristics [2]. The global methods provide the most accurate results but suffer from computational complexity as they require solving the optimisation problem for the entire frame that makes them inappropriate for real-time frame processing.

2.1.1.3 Local Methods

Local methods estimate the optical flow for each pixel. One of the most popular local method for optical flow estimation is the Lucas-Kanade algorithm [3]. The following assumptions are considered:

1. **Colour consistency**. This is the same assumption as for the global methods: a pixel does not change its colour while moving.
2. **Small displacement**. A pixel has a small displacement from one frame to another.

As the pixel has the small displacement it is possible to apply the Taylor series expansion for $I_2(\check{i} + u, \check{j} + v)$ at the point $(\check{i}, \check{j})$ (for notational simplicity subscripts of u and v are omitted, in the context of local methods the optical flow of the current pixel at the point $(\check{i}, \check{j})$ is always assumed):

$$I_2(\check{i} + u, \check{j} + v) \approx I_2(\check{i}, \check{j}) + \frac{\partial I_2}{\partial \check{i}}(\check{i}, \check{j}) \cdot u + \frac{\partial I_2}{\partial \check{j}}(\check{i}, \check{j}) \cdot v \tag{2.2}$$

The colour consistency assumption can be written as follows:

$$I_2(\check{i} + u, \check{j} + v) - I_1(\check{i}, \check{j}) = 0 \tag{2.3}$$

Combination of both assumptions (2.2) and (2.3) results in:

$$I_2(\check{i} + u, \check{j} + v) - I_1(\check{i}, \check{j}) \approx \underbrace{(I_2(\check{i}, \check{j}) - I_1(\check{i}, \check{j}))}_{I_t(\check{i},\check{j})} + \underbrace{\frac{\partial I_2}{\partial \check{i}}}_{I_{\check{i}}}(\check{i}, \check{j}) \cdot u + \underbrace{\frac{\partial I_2}{\partial \check{j}}}_{I_{\check{j}}}(i, j) \cdot v = 0 \tag{2.4}$$

This is an underdetermined problem, because there is one equation and two unknown variables, therefore more assumptions are required.

3. **Spatial similarity**. This is the same assumption as for global methods: neighbouring pixels have the same optical flow.

According to the spatial similarity assumption, each surrounding pixel $(\tilde{i}, \tilde{j})$ from the window $P_{\check{i},\check{j}}$ of the size $\check{K} \times \check{L}$ is supposed to have the same optical flow:

$$I_t(\tilde{i}, \tilde{j}) + I_{\tilde{i}}(\tilde{i}, \tilde{j}) \cdot u + I_{\tilde{j}}(\tilde{i}, \tilde{j}) \cdot v = 0 \qquad \forall (\tilde{i}, \tilde{j}) \in P_{\check{i},\check{j}} \tag{2.5}$$

There are now $\check{K} \cdot \check{L}$ equations and still two unknown variables. Therefore, this is an overdetermined problem and it can be solved using the least squares method.

The conditions of the least squares solution being stable lead to the requirement that the Lucas-Kanade method can be applied to find the optical flow only at interest points such as edges and corners [4].

A review of other gradient-based methods for optical flow estimation can be found in [5]. The methods differ in the basic assumptions, types of estimators and types of motion.

The methods for optical flow estimation can be applied for object detection in video [6–12]. The optical flow methods are used on their own and in combination with other approaches to improve the performance of detection.

2.1.1.4 Background Subtraction

Background subtraction is one of the most popular techniques for object detection. The idea is to separate a background model from the whole scene image. Background represents static parts of the frame. Therefore, the result of this separation (or in other words subtraction) is expected to be moving objects.

The general background subtraction problem can be formulated as follows:

Input:

I — the current frame;

$I_{\text{background}}$ — the background model;

Output:

$\{M_1, \ldots, M_{\acute{K}}\}$ — the set of pixel-masks for each object in the frame I, where $\acute{K}$ is the number of the objects

2.1.1.5 Frame Difference

In the simplest case background subtraction can be implemented as frame difference. In this method the background model $I_{\text{background}}$ is a frame with static background (no objects). A foreground binary mask is then calculated as: $I_{\text{foreground}} = \text{abs}\left(I - I_{\text{background}}\right) < c_{\text{THR}}$, where c_{THR} is a threshold. As the obtained mask $I_{\text{foreground}}$ is usually very noisy, filtering and blob extraction are often performed afterwards.

The main advantage of this method is its simplicity and computational efficiency. The similar approach for contour detection represents the difference of successive frames. The disadvantages of these methods include an inability to work with a slightly changing background, a moving background, illumination changes and others.

2.1.1.6 Gaussian Mixture Model

In the frame difference approach background pixels have a fixed value. Changing values can be modelled as samples from a probability distribution. The common approaches are to use a Gaussian distribution [13] or a mixture of Gaussian distributions [14]. The Gaussian mixture model simulates situations when a pixel represents different types of background at different time moments. For example, one pixel can belong to the sky in one frame and to a tree leaf in another.

The parameter learning of the mixture model can be performed by the expectation-maximisation algorithm but it cannot be applied for online learning. Therefore, the approximate fast procedure is proposed in [14]. New and small components of the mixture model are treated as a foreground as they are poorly fitted with the background model.

The Gaussian mixture model for background subtraction is a good example of a compromise between accuracy and computation complexity. The learning procedure is very fast while a background model is adaptive for light changes.

2.1.1.7 Other Methods

The other approaches for background subtraction differ in features used for background modelling, e.g., texture and colour features are used in [15]; in models for background, e.g., nonparametric estimation of a pixel intensity probability distribution is proposed in [16]; in a spatial level of background modelling, e.g., a hierarchical approach including the pixel, region and frame level for background subtraction is considered in [17]. A number of methods are devoted to special problems of the field, for example, shadow removal from foreground detection [18], specular reflection removal [19] and sudden illumination changes [20]. The survey of conventional and most recent methods can be found in [21–23].

Background subtraction is one of the most widely used methods for object detection in video processing. However, there are still challenges in the field [22]: moving background; moving camera; camouflage (i.e., a foreground object is hardly distinguished from a background); "sleeping foreground" (i.e., a foreground object stops for a long period of time); sudden global illumination changes.

2.1.1.8 Detection of Special Classes of Objects

Both motion detection by optical flow estimation and background subtraction aim to detect moving objects. The interest can be in detection of special classes of objects without considering their motion. Face detection [24] is an example of such kind of applications, where the Viola-Jones face detector [25, 26] is one of the most popular algorithms. Pedestrians represent another class of special objects [27]. Recently deep learning methods have been applied for detection of wide classes of predefined objects simultaneously [28–30].

2.1.2 Object Tracking

The next level of understanding and analysing the observed scene after object detection is object tracking. Object tracking is the process of object localisation and association of its location in the current frame with the previous ones, building a track for each object. A track is a sequence of object locations over time.

The problem of object tracking can be formulated as follows:

Input:

$I_1, \ldots, I_J$ — the sequence of frames;

Output:

$\{Tr_a\}_{a=1}^{\acute{K}}$ — the set of tracks, where $\acute{K}$ is the number of all detected objects over the whole frame sequence, $Tr_a = \{x_{a,\tau_1^a}, \ldots, x_{a,\tau_{\acute{N}_a}^a}\}$ is the track for the a-th object, x_{a,τ_n^a} is the location of the a-th object at time τ_n^a, $\acute{N}_a$ is the total number of moments when the a-th object is tracked

The methods for object tracking can be classified into two very broad categories: tracking matching methods and state space models.

2.1.2.1 Tracking Matching Methods

Tracking methods in this category do not model motion of the objects precisely. They may rely on object detection followed by finding correspondence between detections in different frames, or may be based on object appearance.

The former methods, in general, work by the following scheme: for each frame independently some object detection method is applied and association between these detections in different frames is then found. The association problem can be formulated as the assignment problem [31]. A cost function for the assignment problem can be built based on different assumptions, for example, smoothness of object velocities [32], correlation between new detections and current object templates [33], or matching of the current object description with a region that surrounds new detections [34].

Appearance-based methods seek regions in frames that are matching given object templates. One of the basic methods for such types of trackers is the mean-shift tracker [35]. The mean-shift tracker performs the gradient-ascent approach to find a new location of an object maximising similarity with the given template. The task can be considered as a binary classification problem distinguishing the object and the background [36].

2.1.2.2 State Space Models

The other category of tracking methods is based on state space models. These methods estimate an object state (e.g., position, velocity and acceleration) considering a motion model corrected by incomplete measurements. The measurements are obtained by an object detection algorithm.

Let $\mathbf{X}_\tau$ denote the state of the object at the time τ and $\mathbf{Z}_\tau$ the measurement of the object obtained from the frame at time τ. The motion model represents a functional dependence between the current and previous states:

$$\mathbf{X}_\tau = f_\tau^{\text{motion}}(\mathbf{X}_{\tau-1}, \upsilon_{\tau-1}), \tag{2.6}$$

where $f_\tau^{\text{motion}}(\cdot,\cdot)$ is the motion function, $\upsilon_{\tau-1}$ is noise.

The measurement model represents a dependence between the current object measurement and its current state:

$$\mathbf{Z}_\tau = h_\tau^{\text{meas}}(\mathbf{X}_\tau, u_\tau), \tag{2.7}$$

where $h_\tau^{\text{meas}}(\cdot,\cdot)$ is the measurement function, u_τ is noise, assumed to be independent of υ_τ.

Bayesian filters are applied to find a posterior probability distribution $p(\mathbf{X}_\tau|\mathbf{Z}_{1:\tau})$ of the current state $\mathbf{X}_\tau$ given the sequence of the measurements up to the current moment $\mathbf{Z}_{1:\tau} = \{\mathbf{Z}_1, \ldots, \mathbf{Z}_\tau\}$. Assuming that the posterior distribution for the previous state $p(\mathbf{X}_{\tau-1}|\mathbf{Z}_{1:\tau-1})$ is available, a recursive filtering procedure can be formulated as the following two steps:

prediction step

Compute the predictive density of the current state utilizing the motion model for $p(\mathbf{X}_\tau|\mathbf{X}_{\tau-1})$:

$$p(\mathbf{X}_\tau|\mathbf{Z}_{1:\tau-1}) = \int p(\mathbf{X}_\tau|\mathbf{X}_{\tau-1})p(\mathbf{X}_{\tau-1}|\mathbf{Z}_{1:\tau-1})\mathrm{d}\mathbf{X}_{\tau-1}; \tag{2.8}$$

update step

Compute the posterior probability of the current state utilizing the measurement model for $p(\mathbf{Z}_\tau|\mathbf{X}_\tau)$:

$$p(\mathbf{X}_\tau|\mathbf{Z}_{1:\tau}) = \frac{p(\mathbf{Z}_\tau|\mathbf{X}_\tau)p(\mathbf{X}_\tau|\mathbf{Z}_{1:\tau-1})}{\int p(\mathbf{Z}_\tau|\tilde{\mathbf{X}}_\tau)p(\tilde{\mathbf{X}}_\tau|\mathbf{Z}_{1:\tau-1})\mathrm{d}\tilde{\mathbf{X}}_\tau} \tag{2.9}$$

Different further assumptions about the motion and measurement models and noise lead to different methods. If both υ_τ and u_τ are assumed to be independent and having Gaussian distributions, both motion and measurement models are linear with additive noise, the resulting filter is the Kalman filter [37]. The common extensions of the Kalman filter include the extended and unscented Kalman filters [38]. Although they relax the functional linearity constraint they fail to represent multimodal or heavily skewed posterior distributions. In this case the particle filter can be used [39, 40]. The idea of the particle filter is to represent a posterior distribution, which cannot be calculated analytically, with a weighted sum of samples (particles). An overview of particle filtering for tracking in video can be found in [41].

2.2 Anomaly Detection

In recent years a number of methods have been proposed for the anomaly detection problem. One of the challenges in this field is the absence of one formal problem formulation due to the broadness and informality of the desired objectives. The goal is to detect any unusual events that can be of interest for a human operator and to warn about them. Formalisation of the concept of normality can be achieved by answering the following questions:

- How to represent the video data? The raw video data should be expressed in a manner suitable for further analysis and this representation should contain as much of relevant information from the video as possible.
- How to model behaviours and/or activities happening within the observed scene? How to make the system understand and explain activities? The system should have a model of normal behaviour in order to detect abnormal behaviours.
- What should be considered as normal and what as abnormal? How to measure normality? The system should make a decision about normality or abnormality of the given data.

The problem of abnormal behaviour detection can then be formulated as follows:

Input:
$I_1, \ldots, I_J$ — the sequence of frames;
$D(I_1), \ldots, D(I_J)$ — the descriptors of the frames;
Output:
$\text{label}(D(I_j))$ – the normal or abnormal label $\forall j \in \mathbf{J}_{\text{test}}$, where $\mathbf{J}_{\text{test}} \subseteq \{1, \ldots, J\}$

Note that the decision about abnormality is not limited to be made at a frame level. It can be done both more broadly for a sequence of successive frames and more precisely for a region of a frame.

A review of the methods that answer the above questions is presented below.

2.2.1 *Video Representation*

There is a number of different representations used for feature extraction. They can be based on trajectories or pixel level features.

2.2.1.1 Trajectory-Based Methods

This is an intuitive way of a data representation for the anomaly detection task. In most cases abnormal behaviour can be determined when an individual motion differs from regular motion patterns, i.e., when an individual trajectory differs from regular trajectories.

All object tracking methods are suitable for the anomaly detection problem. For example, in [42] tracking of foreground corner points by the Kanade-Lucas-Tomasi algorithm [43] is performed for anomaly detection purposes. For a given time interval, a set of trajectories is collected by the tracking procedure. Two types of visual features are formed as follows. The scene is divided by grid cells. Coordinates of cells that contain any of trajectories form the first type of features. Features of the second type represent information about relative shifts of cell points along trajectories.

In [44] video representation for anomaly detection is performed by the object detection and tracking framework proposed in [18]. Vectors that describe transitions of objects along their trajectories and object size parameters are used as visual features for behaviour modelling.

Background subtraction and tracking based on detections correspondence are implemented in [45]. At every time moment each track is represented as a vector of current object coordinates, object size parameters and a current velocity. To discretise this representation k-means clustering is performed. Sequences of clusters indices for each trajectory are used as visual features.

2.2.1.2 Methods Based on Pixel Level Features

The other authors argue that the tracking performance still suffers in crowded scenes and use pixel-level features.

All kinds of image descriptors can be used as the visual features of the microscopic type. The recent survey of visual descriptors can be found in [46]. In the literature devoted to anomaly detection in video, there are examples of using very different kinds of image descriptors. A raw value of pixel intensity can be treated as a visual feature [47]. In [48] a frame sequence is divided into 3 dimensional cubes. The scale-invariant feature transform, histogram of oriented gradients and histogram of optical flow features are extracted for each voxel. The local features are then clustered to obtain a bag-of-words histogram for each cube.

Discretised optical flow is a widely used low-level feature. One of the discretisation scheme [49–52] is as follows. The scene is divided into small cells. For each cell a mean optical flow is calculated. This mean motion is then quantised into the given number of directions. This quantised motion and cell coordinates together form visual features. The variations of this scheme include preprocessing of the frame with background subtraction such that an optical flow is calculated only for foreground pixels [53, 54] or averaging of the optical flow in a temporal domain among the number of successive frames [55].

The optical flow is also used as a building block for more sophisticated features. Particle advection inspired by fluid dynamics is proposed in [56]. The model considers small particles moving under the optical flow similarly to leaves moving in a water flow. A social force of interaction is then calculated based on the difference between the actual optical flow for the particle and the average optical flow among neighbours of the particle. In [57] an optical flow is averaged in a temporal domain. The desired velocity is defined as the spatial average of the optical flow over a small neighbourhood of the particle position. In [58] the length of the social force vector is calculated according to the distance between two particles while the direction of the force vector is defined by the optical flow difference. The weighted modification of the initial interaction force is considered in [59]. The modification includes geometrical consistency, social disorder and crowd congestion constraints.

2.2.1.3 Dimensionality Reduction

It is common to perform a dimensionality reduction step after extraction of low-level features. This may be done by clustering [60] or topic modeling [51]. The adopted version of the fuzzy clustering algorithm [61] is applied for the dimensionality reduction in [62]. In [63] a minimum subset of initial codewords is computed such that all the other elements can be reconstructed using this optimal subset.

2.2.2 Behaviour Model

The behaviour model is an abstraction used in a system for anomaly detection to interpret activities happening in the scene. The behaviour models can be classified by the learning procedure as supervised, semi-supervised, or unsupervised.

Supervised models

Supervised models for abnormal behaviour detection are suitable for applications where a particular type of abnormality is expected. These models also require a labelled training dataset. An illustrative example of such kind of applications is fall detection [64, 65].

Semi-supervised models

Semi-supervised models represent a compromise between complexity of a fully-automatic system and expense of labelling the data. Systems that employ semi-supervised models in anomaly detection can work by the following scheme [45, 50]: a system makes alerts to a human operator about all types of abnormalities it finds and the human operator can label some outputs as uninteresting or label missed anomalies. The system should take into account the responses by the human operator and improve the output.

Unsupervised models

Unsupervised methods [66] are used in fully autonomous systems where a system cannot accept any responses from a human operator and it only informs him or her about the abnormality. The human operator is assumed to react on the abnormal events, not to participate in anomaly detection. Methods for semi-supervised and unsupervised learning that represent a main interest in this thesis are reviewed in more detail below.

Although the above classification is traditional for machine learning problems another classification of behaviour models seems to be more discriminative according to the literature review. In this classification the behaviour models are divided into two categories: template-based and statistical models. The former contains models where an explicit template for a normal behaviour is built. While the latter contains models where statistical regularities are extracted from the data.

2.2.2.1 Template-Based Models

In this category a method extracts some templates from data and treats them as a normal behaviour model. In [57] the sum of the visual features of a reference frame is treated as a normal behaviour template. Another common approach is clustering of visual features where clusters are considered as normal behaviour templates [58, 62]. The agglomerative clustering is implemented in [45], where a hidden Markov model is built for every cluster.

2.2.2.2 Statistical Models

The statistical behaviour models find statistical regularities in the data to explain different behaviours within the observed scene.

Behaviour modelling can be viewed as a classical classification problem. In the case of unsupervised learning this is a one-class classification problem [67]. This may be solved by the one-class support vector machine [48]. In the case of supervised learning all conventional binary classifiers are suitable. For example, it may be the two-class support vector machine [59, 68].

Another approach is to consider the problem within the probabilistic framework. Histogram approximation of a probability distribution of the most recent observation is employed in [69]. A Gaussian distribution of a cluster of visual features is considered in [55]. Parameters of the Gaussian distribution are approximated with their sample estimates. The Gaussian mixture model simulates a probability distribution of visual features in [44]. The coupled hidden Markov model that represents spatio-temporal motion dependencies is used for behaviour representation in [70].

Typical activities or behaviours can be considered as sets of features that often appear together. Topic modeling is an approach to find such kind of statistical regularities thus a topic model can be used as a behaviour model in anomaly detection [53, 54, 56, 71]. A number of variations of the conventional topic models have been proposed recently. In [49] different video clips are assumed to have similar mixtures of activities, that is modelled by shared hyperparameters. Temporal dependencies among activities are considered in [50–52]. A continuous model for an object velocity is proposed in [42].

2.2.3 Normality Measure

Once the behaviour model is defined a decision rule about abnormality has to be designed. A new observation is labelled as normal or abnormal based on some normality measure.

If normal behaviour templates are available, the anomaly decision rule is based on comparison of the new observation with these templates. In [57] the simple absolute difference between the new observation and the reference normal template is used as the normality measure. When the difference is larger than a threshold the new observation is considered as abnormal. The Jensen-Shannon divergence [72], which is a symmetrised and smoothed version of the Kullback-Leibler divergence, is used as a similarity measure between the new observation and reference ones in [71]. The Z-score value is considered in [45, 58]. The so-called sparse reconstruction cost is proposed as a normality measure in [63]. The idea is that a normal behaviour is well represented in the basis built from the training data and has a sparse coefficient vector in this basis while an abnormal behaviour cannot be explained with the normal templates and has a dense coefficient vector.

In the case of the probabilistic models, such as topic or hidden Markov models, an intuitive way is to use a likelihood function as normality measure [42, 44, 49, 54–56, 70]. In [69] the likelihood of the new observation is calculated in the multiple spatial locations. The integrated decision about abnormality is then made based on the local likelihood measurers. The comparison of the different normality measures based on the likelihood estimation is provided in [53].

Chapters 3 and 4 consider the topic modeling approach for behaviour analysis and anomaly detection. A brief introduction of topic modeling is provided below.

2.3 Topic Modeling

Topic modeling is an approach to discover statistical patterns in the data. It was originally developed for text mining [73, 74] although it was applied later in many areas including computer vision, genetics [75], collaborative filtering [76, 77] and social network analysis [78]. For clarity the following explanation is provided for the text mining problem.

Topic modeling aims to find latent *topics* given the collection of unlabelled text *documents* that consist of *words*. It is assumed that topics should explain the appearance of words in documents forming a hidden structure of the collection. In probabilistic topic modeling the documents are assumed to be represented as mixtures of topics, where each topic is a distribution over words. It is used for more compact document representation, semantic information retrieval [79], analysing evolution of topics over the time [80, 81], text classification [82] and others.

2.3.1 *Problem Formulation*

The simplest probabilistic topic modeling problem [73, 74] can be formulated as follows:

Input:
$\mathcal{J}$ — a set of documents;
$\mathcal{V}$ — a set of words;
$\mathbf{F} = \|n_{wj}\|_{w \in \mathcal{V}, j \in \mathcal{J}}$ — a co-occurrence matrix of words and documents, where n_{wj} is the number of times the word w appears in the document j;
Assume:
$\mathcal{K}$ — a set of topics;
Output:
$p(w|k)$ — a probability of the word in the topic $\forall w \in \mathcal{V}, k \in \mathcal{K}$;
$p(k|j)$ — a probability of the topic in the document $\forall k \in \mathcal{K}, j \in \mathcal{J}$

Here the **bag-of-word** and **bag-of-document assumptions** are applied, i.e., the joint probability of observed data is independent of the order of words in documents and the order of the documents. The **conditional independence assumption** is also usually used, which means that a word appearance in a document depends only on the corresponding topic but not on the document:

$$p(w|k, j) \equiv p(w|k) \tag{2.10}$$

Given the conditional independence assumption and the law of total probability a probability of a word in a document can be described by:

$$p(w|j) = \sum_{k \in \mathcal{K}} p(w|k)p(k|j) \tag{2.11}$$

The following generative process is assumed for each $j \in \mathcal{J}$:

1. choose the length of the document N_j;
2. repeat N_j times:
   (a) draw a topic k from $p(k|j)$;
   (b) draw a word w from $p(w|k)$

The representation (2.11) can be viewed as a stochastic matrix[1] factorisation problem for the given word-in-document frequency matrix $\hat{\mathbf{F}}$:

$$\hat{\mathbf{F}} = \|\acute{n}_{wj}\|_{w \in \mathcal{V}, j \in \mathcal{J}}, \quad \acute{n}_{wj} = \frac{n_{wj}}{N_j} \tag{2.12}$$

The aim is to represent this matrix as a product of two stochastic matrices $\mathbf{\Phi}$ and $\mathbf{\Theta}$

$$\begin{aligned} \mathbf{\Phi} &= \{\phi_{wk}\}_{w \in \mathcal{V}, k \in \mathcal{K}}, & \phi_{wk} &= p(w|k), & \phi_k &= \{\phi_{wk}\}_{w \in \mathcal{V}}; \\ \mathbf{\Theta} &= \{\theta_{kj}\}_{k \in \mathcal{K}, j \in \mathcal{J}}, & \theta_{kj} &= p(k|j), & \boldsymbol{\theta}_j &= \{\theta_{kj}\}_{k \in \mathcal{K}} \end{aligned}$$

It is worth mentioning that the solution for this representation is not unique [83]. Indeed,

$$\hat{\mathbf{F}} = \mathbf{\Phi} \cdot \mathbf{\Theta} = (\mathbf{\Phi}\mathbf{A}) \cdot (\mathbf{A}^{-1}\mathbf{\Theta}) \tag{2.13}$$

for any $\mathbf{A}$ such that matrices $\mathbf{\Phi}\mathbf{A}$ and $\mathbf{A}^{-1}\mathbf{\Theta}$ are stochastic.

2.3.2 *Inference*

There are two basic topic models: probabilistic latent semantic analysis (PLSA) [73] and latent Dirichlet allocation (LDA) [74]. The former considers inference via maximum likelihood estimation while the latter solves the problem within the Bayesian framework.

---

[1] Stochastic matrix is a non-negative matrix with each column summing to one.

2.3.2.1 Probabilistic Latent Semantic Analysis

In PLSA the parameters $\mathbf{\Phi}$ and $\mathbf{\Theta}$ are estimated by the maximum likelihood approach. The full log likelihood of the collection using (2.11) can be written as follows[2]

$$\log(\mathcal{L}) = \log\left(\prod_{j\in\mathcal{J}}\prod_{w\in\mathcal{V}} p(w|j)^{n_{wj}}\right) = \sum_{j\in\mathcal{J}}\sum_{w\in\mathcal{V}} n_{wj}\log\left(\sum_{k\in\mathcal{K}}\phi_{wk}\theta_{kj}\right) \tag{2.14}$$

The expectation-maximisation (EM) algorithm [84] is applied to parameters that maximise the log likelihood. This is an iterative procedure that repeats E and M-steps. During the E-step the expected value of the topic posterior for each word and document pair is approximated, keeping the current estimates of parameters fixed:

$$h_{kjw} \stackrel{\text{def}}{=} p(k|j,w) = \frac{\phi_{wk}\,\theta_{kj}}{\sum\limits_{k'\in\mathcal{K}}\phi_{wk'}\,\theta_{k'j}} \qquad \forall k\in\mathcal{K}, w\in\mathcal{V}, j\in\mathcal{J} \tag{2.15}$$

The M-step finds parameter estimates, that maximise the log likelihood, keeping the approximation of the topic posterior fixed:

$$\hat{\phi}_{wk} = \frac{\sum\limits_{j\in\mathcal{J}} n_{wk}h_{kjw}}{\sum\limits_{w'\in\mathcal{V}}\sum\limits_{j\in\mathcal{J}} n_{w'j}h_{kw'j}} \qquad \forall w\in\mathcal{V}, k\in\mathcal{K}; \tag{2.16}$$

$$\hat{\theta}_{kj} = \frac{\sum\limits_{w\in\mathcal{V}} n_{wj}h_{kjw}}{N_j} \qquad \forall k\in\mathcal{K}, j\in\mathcal{J} \tag{2.17}$$

2.3.2.2 Latent Dirichlet Allocation

In LDA the Dirichlet priors for the parameters $\mathbf{\Phi}$ and $\mathbf{\Theta}$ are considered:

$$\phi_k \sim Dir(\eta) \qquad \forall k\in\mathcal{K}; \tag{2.18}$$

$$\theta_j \sim Dir(\alpha) \qquad \forall j\in\mathcal{J}, \tag{2.19}$$

where $Dir(\cdot)$ denotes a Dirichlet distribution, η and α are the corresponding parameters of the Dirichlet distributions.

The Bayesian approach includes computation of a posterior distribution of latent variables. In the LDA a true posterior of the parameters is intractable. The main inference schemes for the LDA are variational Bayes [74] and Gibbs sampling [85].

[2]The term $p(j)$ is omitted as it is not used in parameter optimisation.

The **variational Bayes** scheme [86] finds the approximation of the true posterior of latent variables in the class of factorised distributions. In the LDA latent variables are the parameters $\mathbf{\Phi}$ and $\mathbf{\Theta}$ and topic assignments z_{ji} for the word i in the document $j, \forall j \in \mathcal{J}, i \in \{1, \ldots, N_j\}$. Let $\mathbf{z} = \{z_{ji}\}_{j \in \mathcal{J}, i \in \{1,\ldots,N_j\}}$ denote a set of all topic assignments.

The true posterior $p(\mathbf{z}, \mathbf{\Phi}, \mathbf{\Theta}|\mathbf{F}, \boldsymbol{\alpha}, \boldsymbol{\eta})$ is then approximated by the factorised distribution:

$$q(\mathbf{z}, \mathbf{\Phi}, \mathbf{\Theta}) = \prod_{j \in \mathcal{J}} \prod_{i=1}^{N_j} q(z_{ji}|\widetilde{\mathbf{h}}_{ji}) \prod_{j \in \mathcal{J}} q(\boldsymbol{\theta}_j|\tilde{\boldsymbol{\alpha}}_j) \prod_{k \in \mathcal{K}} q(\boldsymbol{\phi}_k|\tilde{\boldsymbol{\eta}}_k), \tag{2.20}$$

where $q(z_{ji}|\widetilde{\mathbf{h}}_{ji})$ is the categorical distribution with the parameter vector $\widetilde{\mathbf{h}}_{ji}$, $q(\boldsymbol{\theta}_j|\tilde{\boldsymbol{\alpha}}_j)$ and $q(\boldsymbol{\phi}_k|\tilde{\boldsymbol{\eta}}_k)$ are the Dirichlet distributions with the parameter vectors $\tilde{\boldsymbol{\alpha}}_j$ and $\tilde{\boldsymbol{\eta}}_k$, respectively.

The update formulae for the parameters of the factorised distributions within an iterative optimisation procedure are:

$$\tilde{\alpha}_{kj} = \alpha_k + \sum_{i=1}^{N_j} \widetilde{h}_{kji} \qquad \forall k \in \mathcal{K}, j \in \mathcal{J}; \tag{2.21}$$

$$\tilde{\eta}_{wk} = \eta_w + \sum_{j \in \mathcal{J}} \sum_{i=1}^{N_j} \widetilde{h}_{kji} \mathbb{1}(w_{ji} = w) \qquad \forall w \in \mathcal{V}, k \in \mathcal{K}; \tag{2.22}$$

$$\widetilde{h}_{kji} \propto \exp\left(\psi(\tilde{\alpha}_{kj}) + \psi(\tilde{\eta}_{w_{ji}k}) - \psi\left(\sum_{w \in \mathcal{V}} \tilde{\eta}_{wk}\right)\right) \qquad \forall k \in \mathcal{K}, j \in \mathcal{J}, i \in \{1, \ldots, N_j\}, \tag{2.23}$$

where w_{ji} is the word at the position i in the document j, $\mathbb{1}(\cdot)$ is the indicator function, $\psi(\cdot)$ is the digamma function, $\tilde{\alpha}_{kj}$, α_k, $\widetilde{h}_{kji}$ and $\tilde{\eta}_{wk}$ are elements of the vectors $\tilde{\boldsymbol{\alpha}}_j$, $\boldsymbol{\alpha}$, $\widetilde{\mathbf{h}}_{ji}$ and $\tilde{\boldsymbol{\eta}}_k$, respectively.

After convergence the point estimates for the model parameters $\mathbf{\Phi}$ and $\mathbf{\Theta}$ can be obtained with the expected values of the respective distributions:

$$\hat{\phi}_{wk} = \frac{\tilde{\eta}_{wk}}{\sum_{w' \in \mathcal{V}} \tilde{\eta}_{w'k}} = \frac{\eta_w + \sum_{j \in \mathcal{J}} \sum_{i=1}^{N_j} \tilde{h}_{kji} \mathbb{1}(w_{ji} = w)}{\sum_{w' \in \mathcal{V}} \left(\eta_{w'} + \sum_{j \in \mathcal{J}} \sum_{i=1}^{N_j} \tilde{h}_{kji} \mathbb{1}(w_{ji} = w')\right)} \qquad \forall w \in \mathcal{V}, k \in \mathcal{K}; \tag{2.24}$$

$$\hat{\theta}_{kj} = \frac{\tilde{\alpha}_{kj}}{\sum_{k' \in \mathcal{K}} \tilde{\alpha}_{k'j}} = \frac{\alpha_k + \sum_{i=1}^{N_j} \tilde{h}_{kji}}{\sum_{k' \in \mathcal{K}} \left(\alpha_{k'} + \sum_{i=1}^{N_j} \tilde{h}_{k'ji}\right)} \qquad \forall k \in \mathcal{K}, j \in \mathcal{J} \tag{2.25}$$

The **Gibbs sampling** scheme [87] is an example of the Markov chain Monte Carlo method. In this framework a Markov chain is built, that is used to represent an intractable distribution by samples. In the Gibbs sampling schemes, the Markov chain is obtained by sampling each one-dimensional hidden variable from its conditional distribution given the current values for all the other latent variables. In LDA a collapsed version of Gibbs sampling is employed where only topic assignment variables z_{ji} are sampled and the parameters $\mathbf{\Phi}$ and $\mathbf{\Theta}$ are then estimated from the computed variables z_{ji}.

The Gibbs sampling update for the topic assignment z_{ji} is as follows:

$$z_{ji} \sim p(z_{ji} = k|\mathbf{w}, \mathbf{z}^{-ji}) \propto \frac{l_{w_{ji}k}^{-ji} + \eta_{w_{ji}}}{\sum\limits_{w \in \mathcal{V}} (l_{wk}^{-ji} + \eta_w)} \frac{n_{kj}^{-ji} + \alpha_k}{\sum\limits_{k' \in \mathcal{K}} (n_{k'j}^{-ji} + \alpha_k)}, \tag{2.26}$$

where $\mathbf{w} = \{w_{ji}\}_{j \in \mathcal{J}, i \in \{1,\dots,N_j\}}$ is a set of all the words, l_{wk} denotes the number of times the word w is associated with the topic k, n_{kj} denotes the number of times the topic k is associated with the document j; the superscript $-ji$ denotes the variables or counts excluding those that correspond to the token i in the document j.

The point estimates of the model parameters $\mathbf{\Phi}$ and $\mathbf{\Theta}$ from one posterior sample of $\mathbf{z}$ can be calculated as follows:

$$\hat{\phi}_{wk} = \frac{l_{wk} + \eta_w}{\sum\limits_{w' \in \mathcal{V}} (l_{w'k} + \eta_{w'})} \qquad \forall w \in \mathcal{V}, k \in \mathcal{K}; \tag{2.27}$$

$$\hat{\theta}_{kj} = \frac{n_{kj} + \alpha_k}{\sum\limits_{k' \in \mathcal{K}} \left(n_{k'j} + \alpha_{k'}\right)} \qquad \forall k \in \mathcal{K}, j \in \mathcal{J} \tag{2.28}$$

2.3.3 *Extensions of Conventional Models*

There are a lot of extensions of the conventional PLSA and LDA models, proposed over the last decade. A review of the models including external information such as authors of documents, dates of publications and correlations between documents can be found in [82]. A number of papers is devoted to the sparseness of the target distributions, e.g., [88]. In [83] different forms of regularisation are presented to overcome the problem of non-uniqueness of the matrix factorisation in topic modeling.

2.3.4 *Dynamic Topic Models*

In conventional topic models, the bag of documents assumption is used. They share the same set of topics, but weights in a topic mixture for a particular document are independent of weights for all other documents in a dataset. However, in some classes of applications it is reasonable to assume similarity of topic mixtures in different documents.

Consider the analysis of scientific papers from a conference in text mining. It is expected that if a topic is "hot" in the current year, it would be popular in the next year too. The popularity of the topics changes through the years but in each two successive years the set of popular topics would often be similar. It means that topic mixtures in documents in successive years are similar to each other.

The same ideas are relevant in video processing. Documents are usually defined as short video clips extracted from a whole video sequence. Topics represent some local motion patterns. If the clips are sufficiently short, motions started in a given clip would continue in the next clip. Therefore, it may be expected that topic mixtures in the successive clips would be similar.

Two types of dynamics are considered in the topic modeling literature. In the first type, the dynamics are assumed on topic mixtures in documents [50, 52, 89]. This type of the dynamics is described earlier. In the second type, the dynamics are assumed on the topics themselves [90–92], i.e., the distributions over words change over time. There are also papers where both types of dynamics are considered [93, 94].

2.3.5 Topic Modeling Applied to Video Analytics

In order to apply the topic modeling framework to video a visual document and word should be defined. Usually, a whole video sequence is divided into non-overlapping clips and the clips are treated as documents. Any kind of features extracted from the video (for example, those mentioned in Sect. 2.2.1) can be used as words. The only requirement is that features should be discrete, however, there are attempts to relax this constraint [42].

In video processing, topic models are used for scene recognition, semantic scene segmentation, behaviour understanding and abnormal behaviour detection. For example, image categorisation via supervised learning is implemented in [60]. For each category a topic model is learnt and an image is assigned to a category with the largest likelihood. The similar framework for supervised behaviour recognition is proposed in [95] where a two-level hierarchical topic model is designed. In [96] topic representation of images is used as features for the support vector machine classifier for image categorisation. The image retrieval application of topic modeling is presented in [97]. In [98] simultaneous image labelling and annotation using topic modeling is proposed. Topic modeling is used to weight observations and predictions in a tracking framework in [99].

Topic modeling is a beneficial tool for finding latent statistical structures and dependencies in data. The advantage of the framework is an ability to work with unlabelled data and simplicity such that metadata and prior knowledge can be included to a model.

2.4 Change Point Detection

Behaviour analysis and anomaly detection in video can be considered as a change point detection problem. Chapter 5 presents methods for detecting changes in behaviour within the general framework of change point detection. This section briefly reviews the area of change point detection and its application for anomaly detection in video.

2.4.1 Change Point Detection in Time Series Data

Change point detection methods are aimed to detect breaks in the time series stationary regimes such that before a change the data follows one probability distribution and after a change it starts to follow another distribution.

Input:
$\mathbf{y} = \{y_1, \ldots, y_N\}$ — observations, where N is the number of observations;
Output:
$\tau^* \in \{1, \ldots, N\}$ — the time of a change point.

Change point detection is an important area that is applied in a wide range of fields: anomaly detection [100]; cyber-attack detection [101]; finance and economics [102]; motion segmentation [103, 104]; speech processing, analysis of biomedical signals, adaptive filtering and tracking and other problems in signal processing [105–109].

Although introduced in 1954 the cumulative sum (CUSUM) algorithm [110] stays a very popular method for change point detection. There are a lot of extensions of the original algorithm introduced in the literature [111–114]. The CUSUM algorithm is an example of a stopping rule (or control chart) procedure developed for change point detection [115]. However, the CUSUM algorithm can be derived within a statistical hypothesis testing approach. Statistical hypothesis testing is an active area in change point detection where one can formulate a null hypothesis corresponding to a case that there is no change against an alternative that there is a change. A number of diverse algorithms within this approach rely on different assumptions about a change nature and data distribution [116–119]. Another direction in the change point detection area is the Bayesian approach where a posteriori probability of a change is computed [120–125].

The methods for change point detection can be classified into two categories: offline and online. In the offline settings a whole dataset of observations is required to apply a method for change point detection, e.g., change point locations can be found by minimising a global penalty function [126–128]. In contrast online methods process data sequentially that generate alarms about detected changes, e.g., by applying the full Bayesian inference [129], or within a filtering framework [130], or by comparing dissimilarity between two subsets of data indicating a change time between these two subsets [131]. There are methods, integrating both offline and online approaches [132].

Changes in time series data can be considered as functional switches, i.e., a data stream follows one functional dependence before a change and after the change a data generative model is expressed by another function. From this perspective Gaussian processes represent a promising methodology for change point detection as they are often used as a prior on functions [133–136].

A Gaussian process is a generalisation of a Gaussian distribution into an infinite continuous domain, i.e., functions [137]. By definition a Gaussian process is a stochastic process such that for any finite subset of time indices the corresponding random vector has a multivariate Gaussian distribution. A Gaussian process is characterised by its mean and covariance functions.

Gaussian processes are explored for change point detection in the literature. In [138] a Gaussian process is used to perform a one-step ahead prediction and a control chart is used to detect a change point based on the difference between a prediction and an actual observation. The method is online and it uses the whole historical data to make a prediction. However, inference in the Gaussian process framework faces scalability issues with the growth of the training datasets. The use of all historical data adds an additional difficulty to the ability of the algorithm to adapt to different stationary regimes to detect multiple changes.

Change points as changes in hyperparameters of a Gaussian process prior are considered in [139, 140], where in [140] change locations are incorporated into a covariance function. Both methods work within the Bayesian approach and estimate the posterior probability of a change. Following the Bayesian approach unknown hyperparameters should be marginalised, although the obtained integrals are intractable. Therefore, these integrals need to be approximated: e.g., by a grid-based method or Hamiltonian Monte Carlo [139] and by Bayesian Monte Carlo [140].

The statistical hypothesis testing method for change point detection in Gaussian process data is proposed in [141]. In this work a change is considered in a mean value of observed data. This method is offline and designed for detection a single change point.

2.4.2 Anomaly as Change Point Detection

If a representation of a behaviour in video is a time series, then a behavioural abnormality can be defined as an abrupt change in the corresponding time series. The main research questions are then: how to represent a behaviour as a time series and how to detect changes in this time series.

In [100] the Fourier transform is applied for crowd motion. Characteristic functions are then used for estimation of motion probability distributions. Changes are detected by the CUSUM algorithm. Optical flow histograms are the basis for the behaviour representation in [142]. The idea is that within a normal crowd behaviour the successive histograms should be similar as a general motion in the scene is the same. A change is then detected when a similarity measure between histograms has a low value. The same idea with a more sophisticated behaviour representation is presented in [143]. Abnormal events are often characterised with higher velocities of the objects in comparison to the normal behaviour. In [144] an anomaly is detected if the total velocity of moving particles within the current frame is above a threshold. In [145] an anomaly is defined as a change in the direction of a dominant social force vector.

2.5 Summary

This chapter presents an overview of relevant works. To start an analysis of a video sequence one should first extract informative features from it. An overview of the current state of the art in anomaly detection in video is then provided, followed by a brief introduction to the areas of topic modeling and change point detection as employed in the following chapters for behaviour analysis and anomaly detection in video.

References

1. B.K.P. Horn, B.G. Schunck, Determining optical flow. Artif. Intell. **17**(1–3), 185–203 (1981)
2. D. Sun, S. Roth, M. Black, Secrets of optical flow estimation and their principles, in *Proceedings of the 2010 IEEE Conference on Computer Vision and Pattern Recognition (CVPR)*, pp. 2432–2439, June 2010
3. B.D. Lucas, T. Kanade, An iterative image registration technique with an application to stereo vision, in *Proceedings of the 7th International Joint Conference on Artificial Intelligence*, vol. 2, pp. 674–679, Aug 1981
4. J. Shi and C. Tomasi, Good features to track, in *Proceedings of the 1994 IEEE Computer Society Conference on Computer Vision and Pattern Recognition (CVPR)*, pp. 593–600, June 1994
5. D. Fleet, Y. Weiss, Optical flow estimation, in *Handbook of Mathematical Models in Computer Vision*, (Springer US, 2006), pp. 237–257
6. M. Yokoyama, T. Poggio, A contour-based moving object detection and tracking, in *Proceedings of the 2nd Joint IEEE International Workshop on Visual Surveillance and Performance Evaluation of Tracking and Surveillance*, pp. 271–276, Oct 2005
7. C. Papageorgiou, M. Oren, T. Poggio, A general framework for object detection, in *Proceedings of the 6th International Conference on Computer Vision*, pp. 555–562, Jan 1998
8. N. Dalal, B. Triggs, C. Schmid, Human detection using oriented histograms of flow and appearance, in *Proceedings of the 9th European Conference on Computer Vision*, pp. 428–441, May 2006
9. A. Talukder, L. Matthies, Real-time detection of moving objects from moving vehicles using dense stereo and optical flow, in *Proceeding of the 2004 IEEE/RSJ International Conference on Intelligent Robots and Systems (IROS)*, vol. 4, pp. 3718–3725, Sept 2004
10. R. Cucchiara, C. Grana, M. Piccardi, A. Prati, Detecting moving objects, ghosts, and shadows in video streams. IEEE Trans. Pattern Anal. Mach. Intell. **25**(10), 1337–1342 (2003)

11. A. Iketani, A. Nagai, Y. Kuno, Y. Shirai, Detecting persons on changing background, in *Proceedings of the 14th International Conference on Pattern Recognition*, vol. 1, pp. 74–76, Aug 1998
12. L. Wixson, Detecting salient motion by accumulating directionally-consistent flow. IEEE Trans. Pattern Anal. Mach. Intell. **22**(8), 774–780 (2000)
13. C. Wren, A. Azarbayejani, T. Darrell, A. Pentland, Pfinder: real-time tracking of the human body. IEEE Trans. Pattern Anal. Mach. Intell. **19**(7), 780–785 (1997)
14. C. Stauffer, W. Grimson, Adaptive background mixture models for real-time tracking, in *Proceedings of the 1999 IEEE Computer Society Conference on Computer Vision and Pattern Recognition (CVPR)*, vol. 2, pp. 246–252, June 1999
15. J. Yao, J. Odobez, Multi-layer background subtraction based on color and texture, in *Proceedings of the 2007 IEEE Conference on Computer Vision and Pattern Recognition (CVPR)*, pp. 1–8, June 2007
16. A. Elgammal, D. Harwood, L. Davis, Non-parametric model for background subtraction, in *Proceedings of the 6th European Conference on Computer Vision*, pp. 751–767, July 2000
17. O. Javed, K. Shafique, M. Shah. A hierarchical approach to robust background subtraction using color and gradient information, in *Proceedings of the Workshop on Motion and Video Computing*, pp. 22–27, Dec 2002
18. O. Javed, M. Shah, Tracking and object classification for automated surveillance, in *Proceedings of the 7th European Conference on Computer Vision*, pp. 343–357, May 2002
19. S.-N. Lim, A. Mittal, L. Davis, N. Paragios, Fast illumination-invariant background subtraction using two views: error analysis, sensor placement and applications, in *Proceedings of the 2005 IEEE Computer Society Conference on Computer Vision and Pattern Recognition (CVPR)*, vol. 1, pp. 1071–1078, June 2005
20. K. Toyama, J. Krumm, B. Brumitt, B. Meyers, Wallflower: principles and practice of background maintenance, in *Proceedings of the 7th IEEE International Conference on Computer Vision*, vol. 1, pp. 255–261, Sept 1999
21. M. Piccardi, Background subtraction techniques: a review, in *Proceedings of the 2004 IEEE International Conference on Systems, Man and Cybernetics*, vol. 4, pp. 3099–3104, Oct 2004
22. M. Cristani, M. Farenzena, D. Bloisi, V. Murino, Background subtraction for automated multisensor surveillance: a comprehensive review. EURASIP J. Adv. Signal Process. **2010**(1), 343057 (2010)
23. T. Bouwmans, Traditional and recent approaches in background modeling for foreground detection: an overview. Comput. Sci. Rev. **11–12**, 31–66 (2014)
24. C. Zhang, Z. Zhang, A survey of recent advances in face detection. Technical Report MSR-TR-2010-66, Microsoft Reseach, June 2010
25. P. Viola, M. Jones, Rapid object detection using a boosted cascade of simple features, in *Proceedings of the 2001 IEEE Computer Society Conference on Computer Vision and Pattern Recognition (CVPR)*, vol. 1, pp. I–511–I–518, Dec 2001
26. P. Viola, M.J. Jones, Robust real-time face detection. Int. J. Comput. Vis. **57**(2), 137–154 (2004)
27. M. Enzweiler, D. Gavrila, Monocular pedestrian detection: survey and experiments. IEEE Trans. Pattern Anal. Mach. Intell. **31**(12), 2179–2195 (2009)
28. R. Girshick, J. Donahue, T. Darrell, J. Malik, Rich feature hierarchies for accurate object detection and semantic segmentation, in *Proceedings of the 2014 IEEE Conference on Computer Vision and Pattern Recognition (CVPR)*, pp. 580–587, June 2014
29. D. Erhan, C. Szegedy, A. Toshev, D. Anguelov, Scalable object detection using deep neural networks, in *Proceedings of the 2014 IEEE Conference on Computer Vision and Pattern Recognition (CVPR)*, pp. 2147–2154, June 2014
30. S. Ren, K. He, R. Girshick, J. Sun, Faster R-CNN: towards real-time object detection with region proposal networks, in *Advances in Neural Information Processing Systems (NIPS 2015)*, pp. 91–99, Dec 2015
31. J. Munkres, Algorithms for the assignment and transportation problems. J. Soc. Ind. Appl. Math. **5**(1), 32–38 (1957)

32. I. Sethi, R. Jain, Finding trajectories of feature points in a monocular image sequence. IEEE Trans. Pattern Anal. Mach. Intell. PAMI **9**(1), 56–73 (1987) (January)
33. A. Lipton, H. Fujiyoshi, R. Patil, Moving target classification and tracking from real-time video, in *Proceedings of the 4th IEEE Workshop on Applications of Computer Vision*, pp. 8–14, Oct 1998
34. A. Alahi, P. Vandergheynst, M. Bierlaire, M. Kunt, Cascade of descriptors to detect and track objects across any network of cameras. Comput. Vis. Image Underst. **114**(6), 624–640 (2010)
35. D. Comaniciu, V. Ramesh, P. Meer, Kernel-based object tracking. IEEE Trans. Pattern Anal. Mach. Intell. **25**(5), 564–577 (2003)
36. S. Avidan, Ensemble tracking. IEEE Trans. Pattern Anal. Mach. Intell. **29**(2), 261–271 (2007)
37. R.E. Kalman, A new approach to linear filtering and prediction problems. Trans. ASME J. Basic Eng. **82** (Series D), 35–45 (1960) (March)
38. S. Julier, J. Uhlmann, A new extension of the Kalman filter to nonlinear systems, in *Proceedings of SPIE 3068 Signal Processing, Sensor Fusion, and Target Recognition VI*, pp. 182–193, July 1997
39. N. Gordon, D. Salmond, A. Smith, Novel approach to nonlinear/non-Gaussian Bayesian state estimation. IEEE Proc. F (Radar Signal Process) **140**(2), 107–113 (1993)
40. M. Arulampalam, S. Maskell, N. Gordon, T. Clapp, A tutorial on particle filters for online nonlinear/non-Gaussian Bayesian tracking. IEEE Trans. Signal Process. **50**(2), 174–188 (2002)
41. L. Mihaylova, P. Brasnett, N. Canagarajah, D. Bull, Object tracking by particle filtering techniques in video sequences, in *Advances and Challenges in Multisensor Data and Information*. NATO Security Through Science Series (IOS Press, 2007), pp. 260–268
42. H. Jeong, Y. Yoo, K.M. Yi, J.Y. Choi, Two-stage online inference model for traffic pattern analysis and anomaly detection. Mach. Vis. Appl. **25**(6), 1501–1517 (2014)
43. C. Tomasi, T. Kanade, Detection and tracking of point features. Technical report (International Journal of Computer Vision, 1991)
44. A. Basharat, A. Gritai, M. Shah, Learning object motion patterns for anomaly detection and improved object detection, in *Proceedings of the 2008 IEEE Conference on Computer Vision and Pattern Recognition (CVPR)*, pp. 1–8, June 2008
45. K. Ouivirach, S. Gharti, M.N. Dailey, Incremental behavior modeling and suspicious activity detection. Pattern Recognit. **46**(3), 671–680 (2013)
46. Y. Li, S. Wang, Q. Tian, X. Ding, A survey of recent advances in visual feature detection. Neurocomputing **149**, Part B, 736–751 (2015) (February)
47. K. Greenewald, A. Hero, Detection of anomalous crowd behavior using spatio-temporal multiresolution model and Kronecker sum decompositions. ArXiv e-prints (2014)
48. K.-W. Cheng, Y.-T. Chen, W.-H. Fang, Abnormal crowd behavior detection and localization using maximum sub-sequence search, in *Proceedings of the 4th ACM/IEEE International Workshop on Analysis and Retrieval of Tracked Events and Motion in Imagery Stream*, pp. 49–58, Oct 2013
49. X. Wang, X. Ma, Unsupervised activity perception in crowded and complicated scenes using hierarchical Bayesian models. IEEE Trans. Pattern Anal. Mach. Intell. **31**(3), 539–555 (2009)
50. T. Hospedales, S. Gong, T. Xiang, Video behaviour mining using a dynamic topic model. Int. J. Comput. Vis. **98**(3), 303–323 (2012)
51. J. Varadarajan, R. Emonet, J.-M. Odobez, A sparsity constraint for topic models—application to temporal activity mining, in *NIPS-2010 Workshop on Practical Applications of Sparse Modeling: Open Issues and New Directions*, Dec 2010
52. D. Kuettel, M. Breitenstein, L. Van Gool, V. Ferrari, What's going on? Discovering spatio-temporal dependencies in dynamic scenes, in *Proceedings of the 2010 IEEE Conference on Computer Vision and Pattern Recognition (CVPR)*, pp. 1951–1958, June 2010
53. J. Varadarajan, J. Odobez, Topic models for scene analysis and abnormality detection, in *Proceedings of the 12th IEEE International Conference on Computer Vision Workshops (ICCV Workshops)*, pp. 1338–1345, Sept 2009

54. J. Li, S. Gong, T. Xiang, Global behaviour inference using probabilistic latent semantic analysis, in *Proceedings of the British Machine Vision Conference*, pp. 193–202, Sept 2008
55. A. Feizi, A. Aghagolzadeh, H. Seyedarabi, Using optical flow and spectral clustering for behavior recognition and detection of anomalous behaviors, in *Proceedings of the 2013 8th Iranian Conference on Machine Vision and Image Processing (MVIP)*, pp. 210–213, Sept 2013
56. R. Mehran, A. Oyama, M. Shah, Abnormal crowd behavior detection using social force model, in *Proceedings of the 2009 IEEE Conference on Computer Vision and Pattern Recognition (CVPR)*, pp. 935–942, June 2009
57. R. Raghavendra, A. Del Bue, M. Cristani, V. Murino, Optimizing interaction force for global anomaly detection in crowded scenes, in *Proceedings of the 2011 IEEE International Conference on Computer Vision Workshops (ICCV Workshops)*, pp. 136–143, Nov 2011
58. S.-H. Yen, C.-H. Wang, Abnormal event detection using HOSF, in *Proceedings of the 2013 International Conference on IT Convergence and Security (ICITCS)*, pp. 1–4, Dec 2013
59. Y. Zhang, L. Qin, H. Yao, Q. Huang, Abnormal crowd behavior detection based on social attribute-aware force model, in *Proceedings of the 2012 19th IEEE International Conference on Image Processing (ICIP)*, pp. 2689–2692, Sept 2012
60. J. Yang, S. Zhang, G. Wang, M. Li, Scene and place recognition using a hierarchical latent topic model. Neurocomputing **148**, 578–586 (2015)
61. P. Hore, L. Hall, D. Goldgof, Y. Gu, A. Maudsley, A. Darkazanli, A scalable framework for segmenting magnetic resonance images. J. Signal Process. Syst. **54**(1–3), 183–203 (2009)
62. M. Roshtkhari, M. Levine. Online dominant and anomalous behavior detection in videos, in *Proceedings of the 2013 IEEE Conference on Computer Vision and Pattern Recognition (CVPR)*, pp. 2611–2618, June 2013
63. Y. Cong, J. Yuan, J. Liu, Abnormal event detection in crowded scenes using sparse representation. Pattern Recognit. **46**(7), 1851–1864 (2013)
64. N. Noury, A. Fleury, P. Rumeau, A. Bourke, G. Laighin, V. Rialle, J. Lundy, Fall detection-principles and methods, in *Proceedings of the 29th Annual International Conference of the IEEE on Engineering in Medicine and Biology Society*, pp. 1663–1666, Aug 2007
65. M. Mubashir, L. Shao, L. Seed, A survey on fall detection: Principles and approaches. Neurocomputing **100**, 144–152 (2013) (Special issue: Behaviours in video)
66. O. Popoola, K. Wang, Video-based abnormal human behavior recognition-a review. IEEE Trans. Syst. Man Cybern. Part C Appl. Rev. **42**(6), 865–878 (2012)
67. D.M.J. Tax, *One-class classification: Concept learning in the absence of counter-examples*. PhD thesis, Technische Universiteit Delft, 2001
68. A. Iscen, A. Armagan, P. Duygulu, What is usual in unusual videos? Trajectory snippet histograms for discovering unusualness, in *Proceedings of the 2014 IEEE Conference on Computer Vision and Pattern Recognition Workshops (CVPRW)*, pp. 808–813, June 2014
69. A. Adam, E. Rivlin, I. Shimshoni, D. Reinitz, Robust real-time unusual event detection using multiple fixed-location monitors. IEEE Trans. Pattern Anal. Mach. Intell. **30**(3), 555–560 (2008)
70. L. Kratz, K. Nishino, Anomaly detection in extremely crowded scenes using spatio-temporal motion pattern models, in *Proceedings of the IEEE Conference on Computer Vision and Pattern Recognition (CVPR)*, pp. 1446–1453, June 2009
71. Z. Su, H. Wei, S. Wei, Crowd event perception based on spatiotemporal Weber field. J. Electr. Comput. Eng. (2014)
72. J. Lin, Divergence measures based on the Shannon entropy. IEEE Trans. Inf. Theory **37**(1), 145–151 (1991)
73. T. Hofmann, Probabilistic latent semantic indexing, in *Proceedings of the 22nd Annual International ACM SIGIR Conference on Research and Development in Information Retrieval*, pp. 50–57, Aug 1999
74. D.M. Blei, A.Y. Ng, M.I. Jordan, Latent Dirichlet allocation. J. Mach. Learn. Res. **3**, 993–1022 (2003)
75. D.M. Blei, Probabilistic topic models. Commun. ACM **55**(4), 77–84 (2012)

76. T. Hofmann, Collaborative filtering via Gaussian probabilistic latent semantic analysis, in *Proceedings of the 26th Annual International ACM SIGIR Conference on Research and Development in Informaion Retrieval*, pp. 259–266, Aug 2003
77. B. Marlin, Modeling user rating profiles for collaborative filtering, in *Advances in Neural Information Processing Systems (NIPS 2003)*, pp. 627–634, Dec 2003
78. J. Tang, J. Sun, C. Wang, Z. Yang, Social influence analysis in large-scale networks, in *Proceedings of the 15th ACM SIGKDD International Conference on Knowledge Discovery and Data Mining*, pp. 807–816, July 2009
79. E. Yan, Y. Ding, S. Milojevic, C.R. Sugimoto, Topics in dynamic research communities: an exploratory study for the field of information retrieval. J. Inf. **6**(1), 140–153 (2012)
80. J. Zhang, Y. Song, C. Zhang, S. Liu, Evolutionary hierarchical Dirichlet processes for multiple correlated time-varying corpora, in *Proceedings of the 16th ACM SIGKDD International Conference on Knowledge Discovery and Data Mining, KDD'10*, pp. 1079–1088, July 2010
81. A. Chaney, D. Blei, Visualizing topic models, in *Proceedings of the International AAAI Conference on Social Media and Weblogs*, pp. 419–422, June 2012
82. A. Daud, J. Li, L. Zhou, F. Muhammad, Knowledge discovery through directed probabilistic topic models: a survey. Front. Comput. Sci. China **4**(2), 280–301 (2010)
83. K. Vorontsov, A. Potapenko, Additive regularization of topic models. Mach. Learn. **101**(1), 1–21 (2015)
84. A.P. Dempster, N.M. Laird, D.B. Rubin, Maximum likelihood from incomplete data via the EM algorithm. J. R. Stat. Soc. Ser. B (Methodol.) **39**(1), 1–38 (1977)
85. T.L. Griffiths, M. Steyvers, Finding scientific topics. Proc. Natl. Acad. Sci. **101**(suppl 1), 5228–5235 (2004)
86. M. Jordan, Z. Ghahramani, T. Jaakkola, L. Saul, An introduction to variational methods for graphical models. Mach. Learn. **37**(2), 183–233 (1999)
87. S. Geman, D. Geman, Stochastic relaxation, Gibbs distributions, and the Bayesian restoration of images. IEEE Trans. Pattern Anal. Mach. Intell. , PAMI **6**(6), 721–741 (1984)
88. J.-T. Chien, Y.-L. Chang, Bayesian sparse topic model. J. Signal Process. Syst. **74**(3), 375–389 (2014)
89. I. Pruteanu-Malinici, L. Ren, J. Paisley, E. Wang, L. Carin, Hierarchical Bayesian modeling of topics in time-stamped documents. IEEE Trans. Pattern Anal. Mach. Intell. **32**(6), 996–1011 (2010)
90. C. Wang, D. Blei, D. Heckerman, Continuous time dynamic topic models, in *Proceedings of the 24th Annual Conference on Uncertainty in Artificial Intelligence (UAI-08)*, pp. 579–586, July 2008
91. X. Fu, J. Li, K. Yang, L. Cui, L. Yang, Dynamic online HDP model for discovering evolutionary topics from Chinese social texts. Neurocomputing **171**, 412–424 (2016)
92. C. Chen, N. Ding, W. Buntine, Dependent hierarchical normalized random measures for dynamic topic modeling, in *Proceedings of the 29th International Conference on Machine Learning*, pp. 895–902, July 2012
93. D.M. Blei, J.D. Lafferty, Dynamic topic models, in *Proceedings of the 23rd International Conference on Machine Learning*, pp. 113–120, June 2006
94. A. Ahmed, E. Xing, Timeline: a dynamic hierarchical Dirichlet process model for recovering birth/death and evolution of topics in text stream, in *Proceedings of the 26th Annual Conference on Uncertainty in Artificial Intelligence (UAI 2010)*, pp. 20–29, July 2010
95. H. Li, F. Zhang, S. Zhang, Multi-feature hierarchical topic models for human behavior recognition. Sci. China Inf. Sci. **57**(9), 1–15 (2014)
96. M. Zang, D. Wen, K. Wang, T. Liu, W. Song, A novel topic feature for image scene classification. Neurocomputing **148**, 467–476 (2015)
97. E. Hörster, R. Lienhart, W. Effelsberg, B. Möller, Topic models for image retrieval on large-scale databases. ACM Sigmultimedia Rec. **1**(4), 15–16 (2009)
98. C. Wang, D. Blei, F.-F. Li. Simultaneous image classification and annotation, in *Proceedings of the 2009 IEEE Conference on Computer Vision and Pattern Recognition (CVPR)*, pp. 1903–1910, June 2009

99. M. Rodriguez, S. Ali, T. Kanade, Tracking in unstructured crowded scenes, in *Proceedings of the 2009 IEEE 12th International Conference on Computer Vision*, pp. 1389–1396, Sept 2009
100. A. Briassouli, I. Kompatsiaris, Spatiotemporally localized new event detection in crowds, in *Proceedings of the 2011 IEEE International Conference on Computer Vision Workshops (ICCV Workshops)*, pp. 928–933, Nov 2011
101. H. Wang, D. Zhang, K.G. Shin, Change-point monitoring for the detection of dos attacks. IEEE Trans. Dependable Secur. Comput. **1**(4), 193–208 (2004)
102. X. Liu, X. Wu, H. Wang, R. Zhang, J. Bailey, K. Ramamohanarao, Mining distribution change in stock order streams, in *Proceeding of the 2010 IEEE 26th International Conference on Data Engineering (ICDE)*, pp. 105–108, March 2010
103. F. Zhou, F.D.l Torre, J.K. Hodgins, Hierarchical aligned cluster analysis for temporal clustering of human motion. IEEE Trans. Pattern Anal. Mach. Intell. **35**(3), 582–596 (2013)
104. D. Gong, G. Medioni, X. Zhao, Structured time series analysis for human action segmentation and recognition. IEEE Trans. Pattern Anal. Mach. Intell. **36**(7), 1414–1427 (2014)
105. F. Gustafsson, *Adaptive Filtering and Change Detection*, vol. 1. (Wiley, 2000)
106. T.L. Lai, Sequential changepoint detection in quality control and dynamical systems. J. R. Stat. Soc. Ser. B (Methodol.) **57**(4), 613–658 (1995)
107. L. Wang, T. Tan, H. Ning, W. Hu, Silhouette analysis-based gait recognition for human identification. IEEE Trans. Pattern Anal. Mach. Intell. **25**(12), 1505–1518 (2003)
108. F. Han, Z. Tu, S.-C. Zhu, Range image segmentation by an effective jump-diffusion method. IEEE Trans. Pattern Anal. Mach. Intell. **26**(9), 1138–1153 (2004)
109. M. Davy, S. Godsill, Detection of abrupt spectral changes using support vector machines an application to audio signal segmentation, in *IEEE International Conference on Acoustics, Speech, and Signal Processing (ICASSP)*, vol. 2, pp. 1313–1316, May 2002
110. E. Page, Continuous inspection schemes. Biometrika **41**(1/2), 100–115 (1954)
111. W.H. Woodall, M.M. Ncube, Multivariate CUSUM quality-control procedures. Technometrics **27**(3), 285–292 (1985)
112. J.M. Lucas, Combined Shewhart-CUSUM quality control schemes. J. Quality Technol. **14**(2), 51–59 (1982)
113. Y.-K. Lee, D.J. Biau, B.-H. Yoon, T.-Y. Kim, Y.-C. Ha, K.-H. Koo, Learning curve of acetabular cup positioning in total hip arthroplasty using a cumulative summation test for learning curve (LC-CUSUM). J. Arthroplast. **29**(3), 586–589 (2014)
114. A.G. Tartakovsky, B.L. Rozovskii, R.B. Blazek, H. Kim, A novel approach to detection of intrusions in computer networks via adaptive sequential and batch-sequential change-point detection methods. IEEE Trans. Signal Process. **54**(9), 3372–3382 (2006)
115. S.W. Roberts, A comparison of some control chart procedures. Technometrics **8**(3), 411–430 (1966)
116. A. Tartakovsky, I. Nikiforov, M. Basseville, *Sequential analysis: hypothesis testing and changepoint detection* (CRC Press, 2014)
117. S. Teng, Y. Chen, K. Cheng, H. Lo, Hypothesis-test-based landcover change detection using multi-temporal satellite images-a comparative study. Adv. Space Res. **41**(11), 1744–1754 (2008)
118. T.L. Lai, Sequential multiple hypothesis testing and efficient fault detection-isolation in stochastic systems. IEEE Trans. Inf. Theory **46**(2), 595–608 (2000)
119. S.S. Ho, H. Wechsler, A martingale framework for detecting changes in data streams by testing exchangeability. IEEE Trans. Pattern Anal. Mach. Intell. **32**(12), 2113–2127 (2010)
120. A. Shiryaev, The problem of the most rapid detection of a disturbance in a stationary process. Soviet Math. Dokl. **2**, 795–799 (1961)
121. D. Barry, J.A. Hartigan, A Bayesian analysis for change point problems. J. Am. Stat. Assoc. **88**(421), 309–319 (1993)
122. B.P. Carlin, A.E. Gelfand, A.F.M. Smith, Hierarchical Bayesian analysis of changepoint problems. J. R. Stat. Soc. Ser. C (Appl. Stat.) **41**(2), 389–405 (1992)

123. M. Zhitlukhin, A. Shiryaev, Bayesian disorder problems on filtered probability spaces. Theory Probab. Appl. **57**(3), 497–511 (2013)
124. P. Fearnhead, Exact Bayesian curve fitting and signal segmentation. IEEE Trans. Signal Process. **53**(6), 2160–2166 (2005)
125. E. Punskaya, C. Andrieu, A. Doucet, W.J. Fitzgerald, Bayesian curve fitting using MCMC with applications to signal segmentation. IEEE Trans. Signal Process. **50**(3), 747–758 (2002)
126. M. Lavielle, Using penalized contrasts for the change-point problem. Signal Process. **85**(8), 1501–1510 (2005)
127. E. Lebarbier, Detecting multiple change-points in the mean of Gaussian process by model selection. Signal Process. **85**(4), 717–736 (2005)
128. J. Bai, Estimation of a change point in multiple regression models. Rev. Econ. Stat. **79**(4), 551–563 (1997)
129. R. Adams, D. MacKay, Bayesian online changepoint detection. Technical report (University of Cambridge, 2007)
130. B. Azimi-Sadjadi, P. Krishnaprasad, Change detection for nonlinear systems; a particle filtering approach, in *Proceedings of the 2002 American Control Conference*, vol. 5, pp. 4074–4079, May 2002
131. F. Desobry, M. Davy, C. Doncarli, An online kernel change detection algorithm. IEEE Trans. Signal Process. **53**(8), 2961–2974 (2005)
132. C. Alippi, G. Boracchi, M. Roveri, Hierarchical change-detection tests. IEEE Trans. Neural Netw. Learn. Syst. **28**(2), 246–258 (2017)
133. H.-C. Kim, Z. Ghahramani, Bayesian Gaussian process classification with the EM-EP algorithm. IEEE Trans. Pattern Anal. Mach. Intell. **28**(12), 1948–1959 (2006)
134. C.K.I. Williams, D. Barber, Bayesian classification with Gaussian processes. IEEE Trans. Pattern Anal. Mach. Intell. **20**(12), 1342–1351 (1998)
135. M. Filippone, M. Girolami, Pseudo-marginal Bayesian inference for Gaussian processes. IEEE Trans. Pattern Anal. Mach. Intell. **36**(11), 2214–2226 (2014)
136. S. Bratieres, N. Quadrianto, Z. Ghahramani, GPstruct: Bayesian structured prediction using Gaussian processes. IEEE Trans. Pattern Anal. Mach. Intell. **37**(7), 1514–1520 (2015)
137. C.E. Rasmussen, C.K.I. Williams, *Gaussian Processes for Machine Learning* (The MIT Press, 2006)
138. V. Chandola, R.R. Vatsavai, A Gaussian process based online change detection algorithm for monitoring periodic time series, in *Proceedings of the 2011 SIAM International Conference on Data Mining*, pp. 95–106, Apr 2011
139. Y. Saatçi, R.D. Turner, C.E. Rasmussen, Gaussian process change point models, in *Proceedings of the 27th International Conference on Machine Learning*, pp. 927–934, June 2010
140. R. Garnett, M.A. Osborne, S.J. Roberts, Sequential Bayesian prediction in the presence of changepoints, in *Proceedings of the 26th Annual International Conference on Machine Learning*, pp. 345–352, June 2009
141. H. Keshavarz, C. Scott, X. Nguyen, Optimal change point detection in Gaussian processes (2015), arXiv:1506.01338
142. I.R.D. Almeida, C.R. Jung, Change detection in human crowds, in *Proceedings of the 2013 XXVI Conference on Graphics, Patterns and Images*, pp. 63–69, Aug 2013
143. H. Fradi, J.L. Dugelay, Sparse feature tracking for crowd change detection and event recognition, in *Proceedings of the 2014 International Conference on Pattern Recognition (ICPR)*, pp. 4116–4121, Aug 2014
144. V. Kaltsa, A. Briassouli, I. Kompatsiaris, M.G. Strintzis, Timely, robust crowd event characterization, in *Proceedings of the 2012 19th IEEE International Conference on Image Processing*, pp. 2697–2700, Sept 2012
145. D.-Y. Chen, P.-C. Huang, Motion-based unusual event detection in human crowds. J. Vis. Commun. Image Represent. **22**(2), 178–186 (2011)

Chapter 3
Proposed Learning Algorithms for Markov Clustering Topic Model

This chapter introduces the methods for the behaviour analysis and anomaly detection in video using a topic model. Topics in video applications represent typical motion patterns in an observed scene. These patterns can be used for semantic understanding of the typical activities happening within the scene. They can also be used to detect abnormal events. Likelihood of newly observed data is employed as a measure of normality. If something atypical happens in a new visual document, then this document cannot be fitted with the topics, or typical activities, learnt before, and it would have a low likelihood value.

The focus of this chapter is on development and comparison of learning algorithms for the Markov clustering topic model. A novel anomaly localisation procedure is also introduced in this chapter.

The results of the work presented in this chapter are disseminated in:

- O. Isupova, D. Kuzin, L. Mihaylova. "Learning Methods for Dynamic Topic Modeling in Automated Behaviour Analysis", in *IEEE Transactions on Neural Networks and Learning Systems*, vol. PP, no. 99, pp. 1–14
- O. Isupova, L. Mihaylova, D. Kuzin, G. Markarian, F. Septier. "An Expectation Maximisation Algorithm for Behaviour Analysis in Video", in *Proceedings of 18th International Conference on Information Fusion*, 6–9 July 2015, Washington, USA, pp. 126–133
- O. Isupova, D. Kuzin, L. Mihaylova."Abnormal Behaviour Detection in Video Using Topic Modeling", in *Proceedings of University of Sheffield Engineering Symposium*, 24 June 2015, Sheffield, UK

The rest of the chapter is organised as follows. Section 3.1 describes the overall structure of visual documents and visual words. Section 3.2 introduces the dynamic topic model. The new learning algorithms are presented in Sect. 3.3. The methods are given with a detailed discussion about their similarities and differences. The anomaly detection procedure is presented in Sect. 3.4. The learning algorithms are evaluated with real data in Sects. 3.5 and 3.6 summarises the chapter.

O. Isupova, *Machine Learning Methods for Behaviour Analysis and Anomaly Detection in Video*, Springer Theses, https://doi.org/10.1007/978-3-319-75508-3_3

3.1 Video Representation

In order to apply the topic modeling approach to video processing it is required to define visual words and visual documents. In this thesis a visual word is defined as a quantised local motion measured by an optical flow [1–4]. The optical flow vector is discretised spatially by averaging among $\bar{N} \times \bar{N}$ pixels. Those pixel cells that have an average optical flow that exceeds a threshold are called moving cells. The direction of the average optical flow vector for moving cells is further quantised into the four main categories—up, right, down and left (Fig. 3.1). The location of a moving cell and its categorised motion direction together form a visual word.

The whole video sequence is divided into non-overlapping clips. Each clip is a visual document. The document consists of all the visual words extracted from the frames that form the corresponding clip.

Topics in topic modeling are defined as distributions over words. They indicate which words appear together. In the video processing applications topics are distributions over visual words. As visual words represent local motions, topics indicate the set of local motions that frequently appear together. They are usually called *activities* or *actions* (e.g., [1, 2, 5]).

Once visual documents, words and topics are defined, a topic model for video processing can be formulated.

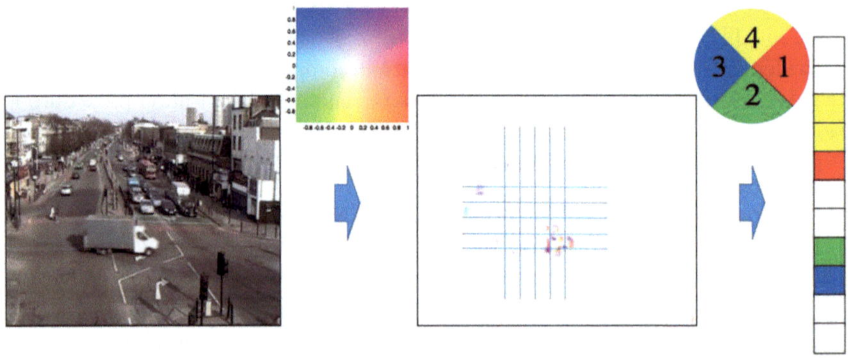

Fig. 3.1 Structure of visual feature extraction: from an input frame (on the left) a map of local motions based on optical flow estimation is calculated (in the centre). For visualisation purposes the calculated optical flow is depicted by the presented colour code. The motion is then quantised into four directions to get the feature representation (on the right)

3.2 Model

3.2.1 Motivation

In topic modeling there are two main kinds of distributions—the distributions over words, which correspond to topics, and the distributions over topics, which characterise the documents. The relationship between documents and words is then represented via latent low-dimensional entities called topics. Having only an unlabelled collection of documents, topic modeling methods restore a hidden structure of the data, i.e., the distributions over words and the distributions over topics.

Consider a set of distributions over topics and a topic distribution for each document is chosen from this set. If the cardinality of the set of distributions over topics is less than the number of documents then documents are clustered into groups, having the same topic distribution within a group. A unique distribution over topics is called a *behaviour* in this work. Therefore, each document corresponds to one behaviour. In topic modeling a document is fully described by a corresponding distribution over topics, which means in this case a document is fully described by a corresponding behaviour.

There are a number of applications where we can observe documents clustered into groups with the same distribution over topics. Let us consider some examples from video analytics where a visual word corresponds to a motion within a tiny cell. As topics represent words that statistically often appear together, in video analytics applications topics define some motion patterns in local areas.

Let us consider a road junction regulated by traffic lights. A general motion on the junction is the same with the same traffic light regime. Therefore, the documents associated with the same traffic light regimes have the same distributions over topics, i.e., they correspond to the same behaviours.

Another example is a video stream generated by a CCTV camera from a train station. Here it is also possible to distinguish several types of general motion within the camera scene: getting off and on a train and waiting for it. These types of motion correspond to behaviours, where the different visual documents showing different instances of the same behaviour have very similar motion structures, i.e., the same topic distribution.

Each action in real life lasts for some time, e.g., a traffic light regime stays the same and people get on and off a train for several seconds. Moreover, often these different types of motion or behaviours follow a cycle and their changes occur in some order. These insights motivate modelling of a sequence of behaviours as a Markov chain, so that the behaviours remain the same during some documents and change in the predefined order. The model that has these described properties is called a Markov Clustering Topic Model (MCTM) in [2]. The next section formally formulates the model.

3.2.2 Model Formulation

This section starts from the introduction of the main notations. Recall that $\mathcal{V}$ is the vocabulary of words and $\mathcal{K}$ is the set of all topics. Denote by $\mathcal{B}$ the set of all behaviours, and b is used to denote an element from this set.

Let $\mathbf{w}_j = \{w_{ij}\}_{i=1}^{N_j}$ denote the set of words for the document j, where N_j is the length of the document j. Let $\mathbf{w}_{1:J_{tr}} = \{\mathbf{w}_j\}_{j=1}^{J_{tr}}$ denote the set of all words for the whole dataset, where J_{tr} is the number of documents in the dataset. Similarly, denote by $\mathbf{z}_j = \{z_{ji}\}_{i=1}^{N_j}$ and $\mathbf{z}_{1:J_{tr}} = \{\mathbf{z}_j\}_{j=1}^{J_{tr}}$ a set of topics for the document j and the set of all topics for the whole dataset, respectively. Let $\mathbf{b}_{1:J_{tr}} = \{b_j\}_{j=1}^{J_{tr}}$ denote a set of all behaviours for all documents.

Note that w and b without subscript denote possible values for a word and behaviour from $\mathcal{V}$ and $\mathcal{B}$, respectively, while the symbols with subscripts denote word and behaviour assignments in particular places in a dataset.

Recall that $\mathbf{\Phi}$ denotes a matrix corresponding to the distributions over words given the topics, $\mathbf{\Theta}$ denotes a matrix corresponding to the distributions over topics given behaviours. For a Markov chain of behaviours a vector $\boldsymbol{\omega}$ for a behaviour distribution for the first document and a matrix $\mathbf{\Xi}$ for transition probability distributions between the behaviours are introduced:

$$
\begin{aligned}
&\mathbf{\Phi} = \{\phi_{wk}\}_{w\in\mathcal{V},k\in\mathcal{K}}, && \phi_{wk} = p(w|k), && \boldsymbol{\phi}_k = \{\phi_{wk}\}_{w\in\mathcal{V}};\\
&\mathbf{\Theta} = \{\theta_{kb}\}_{k\in\mathcal{K},b\in\mathcal{B}}, && \theta_{kb} = p(k|b), && \boldsymbol{\theta}_b = \{\theta_{kb}\}_{k\in\mathcal{K}};\\
&\boldsymbol{\omega} = \{\omega_b\}_{b\in\mathcal{B}}, && \omega_b = p(b); && \\
&\mathbf{\Xi} = \{\xi_{b'b}\}_{b'\in\mathcal{B},b\in\mathcal{B}}, && \xi_{b'b} = p(b'|b), && \boldsymbol{\xi}_b = \{\xi_{b'b}\}_{b'\in\mathcal{B}},
\end{aligned}
$$

where the matrices $\mathbf{\Phi}$, $\mathbf{\Theta}$ and $\mathbf{\Xi}$ and the vector $\boldsymbol{\omega}$ are formed as follows. An element of a matrix on the i-th row and i'-th column is the probability of the i-th element given the i'-th one, e.g., ϕ_{wk} is a probability of the word w in the topic k. The columns of the matrices then form distributions for corresponding elements, e.g., $\boldsymbol{\theta}_b$ is a distribution over topics for the behaviour b. Elements of the vector $\boldsymbol{\omega}$ are probabilities of behaviours to be chosen by the first document. All these distributions are categorical.

The introduced distributions form a set

$$
\mathbf{\Omega} = \{\mathbf{\Phi}, \mathbf{\Theta}, \boldsymbol{\omega}, \mathbf{\Xi}\} \tag{3.1}
$$

of model parameters and they are estimated during a learning procedure.

Prior distributions are imposed to all the parameters. Conjugate Dirichlet distributions are used:

Algorithm 3.2.1 The generative process for the MCTM

Require: The number of clips – J_{tr}, the length of each clip – N_j $\forall j = \{1, \dots, J_{tr}\}$, the hyperparameters – $\boldsymbol{\eta}$, $\boldsymbol{\alpha}$, $\boldsymbol{\varkappa}$, $\boldsymbol{\upsilon}$;
Ensure: The dataset $\mathbf{w}_{1:J_{tr}} = \{w_{11}, \dots, w_{ji}, \dots, w_{J_{tr} N_{J_{tr}}}\}$;
for all $k \in \mathcal{K}$ **do**
  draw a word distribution for the topic k: $\boldsymbol{\phi}_k \sim Dir(\boldsymbol{\eta})$;
for all $b \in \mathcal{B}$ **do**
  draw a topic distribution for behaviour b: $\boldsymbol{\theta}_b \sim Dir(\boldsymbol{\alpha})$;
  draw a transition distribution for behaviour b: $\boldsymbol{\xi}_b \sim Dir(\boldsymbol{\upsilon})$;
draw a behaviour probability distribution for the initial document: $\boldsymbol{\omega} \sim Dir(\boldsymbol{\varkappa})$;
for all $j \in \{1, \dots, J_{tr}\}$ **do**
  if $j = 1$ **then**
    draw a behaviour for the document from the initial distribution: $b_j \sim Cat(\boldsymbol{\omega})$;
  else
    draw a behaviour for the document based on the behaviour of the previous document: $b_j \sim Cat(\boldsymbol{\xi}_{b_{j-1}})$;
  for all $i \in \{1, \dots, N_j\}$ **do**
    draw a topic for the token i based on the chosen behaviour: $z_{ji} \sim Cat(\boldsymbol{\theta}_{b_j})$;
    draw a visual word for the token i based on the chosen topic: $w_{ji} \sim Cat(\boldsymbol{\phi}_{z_{ji}})$;

$$
\begin{aligned}
\boldsymbol{\phi}_k &\sim Dir(\boldsymbol{\phi}_k|\boldsymbol{\eta}), && \forall k \in \mathcal{K}; \\
\boldsymbol{\theta}_b &\sim Dir(\boldsymbol{\theta}_b|\boldsymbol{\alpha}), && \forall b \in \mathcal{B}; \\
\boldsymbol{\omega} &\sim Dir(\boldsymbol{\omega}|\boldsymbol{\varkappa}); && \\
\boldsymbol{\xi}_b &\sim Dir(\boldsymbol{\xi}_b|\boldsymbol{\upsilon}), && \forall b \in \mathcal{B},
\end{aligned}
$$

where $\boldsymbol{\eta}$, $\boldsymbol{\alpha}$, $\boldsymbol{\varkappa}$ and $\boldsymbol{\upsilon}$ are the corresponding hyperparameters of the Dirichlet distributions. As topics and behaviours are not known a priori and will be specified via the learning procedure, it is impossible to distinguish two topics or two behaviours in advance. This is the reason why all the prior distributions are the same for all topics and all behaviours.

The generative process for the model is as follows. All the parameters are drawn from the corresponding prior Dirichlet distributions. At each time j a behaviour b_j is chosen first for a visual document. The behaviour is sampled using the matrix $\boldsymbol{\Xi}$ according to the behaviour chosen for the previous document. For the first document the behaviour is sampled using the vector $\boldsymbol{\omega}$. Once the behaviour is selected, the procedure of choosing visual words repeats N_j times. The procedure consists of two steps—sampling a topic z_{ji} using the matrix $\boldsymbol{\Theta}$ according to the chosen behaviour b_j followed by sampling a word w_{ji} using the matrix $\boldsymbol{\Phi}$ according to the chosen topic z_{ji} for each token $i \in \{1, \dots, N_j\}$, where a token is a particular place inside a document a word is assigned to. The generative process is summarised in Algorithm 3.2.1.[1] The graphical model, showing the relationships between the variables, can be found in Fig. 3.2.

[1] Here, $Cat(\mathbf{v})$ denotes a categorical distribution, where components of a vector $\mathbf{v}$ are probabilities of a discrete random variable to take one of possible values.

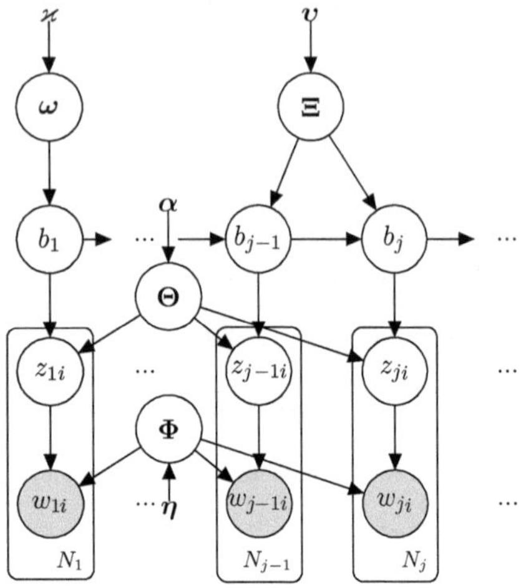

Fig. 3.2 Graphical representation of the Markov Clustering Topic Model

The full likelihood of the observed variables $\mathbf{w}_{1:J_{tr}}$, the hidden variables $\mathbf{z}_{1:J_{tr}}$ and $\mathbf{b}_{1:J_{tr}}$ and the set of parameters $\mathbf{\Omega}$ can be written then as follows:

$$p(\mathbf{w}_{1:J_{tr}}, \mathbf{z}_{1:J_{tr}}, \mathbf{b}_{1:J_{tr}}, \mathbf{\Omega}|\boldsymbol{\eta}, \boldsymbol{\alpha}, \boldsymbol{\varkappa}, \boldsymbol{\upsilon}) = \underbrace{p(\omega|\varkappa)\, p(\Xi|\upsilon)\, p(\mathbf{\Theta}|\alpha)\, p(\mathbf{\Phi}|\eta)}_{\text{Priors}} \times \underbrace{p(b_1|\omega) \left[\prod_{j=2}^{J_{tr}} p(b_j|b_{j-1}, \Xi)\right] \prod_{j=1}^{J_{tr}} \prod_{i=1}^{N_j} p(w_{ji}|z_{ji}, \mathbf{\Phi}) p(z_{ji}|b_j, \mathbf{\Theta})}_{\text{Likelihood}} \tag{3.2}$$

3.3 Parameter Learning

In [2] Gibbs sampling is implemented for parameter learning in the MCTM. We propose two new learning algorithms: based on an EM-algorithm for the maximum a posteriori (MAP) estimates of parameters and based on variational Bayes inference to estimate posterior distributions of the parameters. In this section we introduce the proposed learning algorithms and briefly review the Gibbs sampling scheme.

3.3.1 Expectation-Maximisation Learning

We propose a learning algorithm for MAP estimates of parameters based on the Expectation-Maximisation algorithm [6]. The algorithm consists of repeating E and M-steps. Conventionally, the EM-algorithm is applied to get maximum likelihood estimates (MLE). In that case the M-step is:

$$\mathcal{Q}(\boldsymbol{\Omega}, \boldsymbol{\Omega}^{\text{old}}) \longrightarrow \max_{\boldsymbol{\Omega}}, \tag{3.3}$$

where $\boldsymbol{\Omega}^{\text{old}}$ denotes the set of parameters obtained at the previous iteration and $\mathcal{Q}(\boldsymbol{\Omega}, \boldsymbol{\Omega}^{\text{old}})$ is the expected logarithm of the full likelihood function of the observed and hidden variables:

$$\mathcal{Q}(\boldsymbol{\Omega}, \boldsymbol{\Omega}^{\text{old}}) = \mathbb{E}_{p(\mathbf{z}_{1:J_{tr}}, \mathbf{b}_{1:J_{tr}} | \mathbf{w}_{1:J_{tr}}, \boldsymbol{\Omega}^{\text{old}})} \log p(\mathbf{w}_{1:J_{tr}}, \mathbf{z}_{1:J_{tr}}, \mathbf{b}_{1:J_{tr}} | \boldsymbol{\Omega}). \tag{3.4}$$

The subscript of the expectation sign means the distribution with respect to which the expectation is calculated. During the E-step the posterior distribution of the hidden variables is estimated given the current estimates of parameters.

Here the EM-algorithm is applied to get MAP estimates instead of traditional MLE. The M-step is modified in this case as:

$$\mathcal{Q}(\boldsymbol{\Omega}, \boldsymbol{\Omega}^{\text{old}}) + \log p(\boldsymbol{\Omega} | \boldsymbol{\eta}, \boldsymbol{\alpha}, \boldsymbol{\varkappa}, \boldsymbol{\upsilon}) \longrightarrow \max_{\boldsymbol{\Omega}}, \tag{3.5}$$

where $p(\boldsymbol{\Omega} | \boldsymbol{\eta}, \boldsymbol{\alpha}, \boldsymbol{\varkappa}, \boldsymbol{\upsilon})$ is the prior distribution of the parameters.

As the hidden variables are discrete, the expectation converts to a sum of all possible values for the whole set of the hidden variables $\{\mathbf{z}_{1:J_{tr}}, \mathbf{b}_{1:J_{tr}}\}$. The substitution of the likelihood expression from Eq. (3.2) into (3.5) allows to marginalise some hidden variables from the sum. The remaining distributions that are required for computing the $\mathcal{Q}$-function are following:

- $p(b_1 = b | \mathbf{w}_{1:J_{tr}}, \boldsymbol{\Omega}^{\text{old}})$—the posterior distribution of a behaviour for the first document;
- $p(b_j = b', b_{j-1} = b | \mathbf{w}_{1:J_{tr}}, \boldsymbol{\Omega}^{\text{old}})$—the posterior distribution of two behaviours for successive documents;
- $p(z_{ji} = k | \mathbf{w}_{1:J_{tr}}, \boldsymbol{\Omega}^{\text{old}})$—the posterior distribution of a topic assignment for a given token;
- $p(z_{ji} = k, b_j = b | \mathbf{w}_{1:J_{tr}}, \boldsymbol{\Omega}^{\text{old}})$—the joint posterior distribution of a topic and behaviour assignments for a given token.

With the fixed current values for these posterior distributions the estimates of parameters that maximise the required functional of the M-step (3.5) can be computed as:

$$\widehat{\phi}_{wk}^{\mathrm{EM}} = \frac{\left(\eta_w + \hat{l}_{wk}^{\mathrm{EM}} - 1\right)_+}{\sum\limits_{w' \in \mathcal{V}} \left(\eta_{w'} + \hat{l}_{w'k}^{\mathrm{EM}} - 1\right)_+}, \qquad \forall w \in \mathcal{V}, k \in \mathcal{K}; \tag{3.6}$$

$$\widehat{\theta}_{kb}^{\mathrm{EM}} = \frac{\left(\alpha_k + \hat{n}_{kb}^{\mathrm{EM}} - 1\right)_+}{\sum\limits_{k' \in \mathcal{K}} \left(\alpha_{k'} + \hat{n}_{k'b}^{\mathrm{EM}} - 1\right)_+}, \qquad \forall k \in \mathcal{K}, b \in \mathcal{B}; \tag{3.7}$$

$$\widehat{\xi}_{b'b}^{\mathrm{EM}} = \frac{\left(\upsilon_{b'} + \hat{n}_{b'b}^{\mathrm{EM}} - 1\right)_+}{\sum\limits_{\tilde{b} \in \mathcal{B}} \left(\upsilon_{\tilde{b}} + \hat{n}_{\tilde{b}b}^{\mathrm{EM}} - 1\right)_+}, \qquad \forall b', b \in \mathcal{B}; \tag{3.8}$$

$$\widehat{\omega}_{b}^{\mathrm{EM}} = \frac{\left(\varkappa_b + \hat{n}_{b}^{\mathrm{EM}} - 1\right)_+}{\sum\limits_{b' \in \mathcal{B}} \left(\varkappa_{b'} + \hat{n}_{b'}^{\mathrm{EM}} - 1\right)_+}, \qquad \forall b \in \mathcal{B}, \tag{3.9}$$

where $(a)_+ \overset{\mathrm{def}}{=} \max(a, 0)$ [7]; η_w, α_z, $\varkappa_b$, and $\upsilon_{b'}$ are the elements of the hyperparameter vectors $\boldsymbol{\eta}$, $\boldsymbol{\alpha}$, $\boldsymbol{\varkappa}$ and $\boldsymbol{\upsilon}$, respectively, and:

- $\hat{l}_{wk}^{\mathrm{EM}} = \sum\limits_{j=1}^{J_{tr}} \sum\limits_{i=1}^{N_j} p(z_{ji} = k | \mathbf{w}_{1:J_{tr}}, \boldsymbol{\Omega}^{\mathrm{old}}) \mathbb{1}(w_{ji} = w)$—the expected number of times, when the word w is associated with the topic k;
- $\hat{n}_{kb}^{\mathrm{EM}} = \sum\limits_{j=1}^{J_{tr}} \sum\limits_{i=1}^{N_j} p(z_{ji} = k, b_j = b | \mathbf{w}_{1:J_{tr}}, \boldsymbol{\Omega}^{\mathrm{old}})$—the expected number of times, when the topic k is associated with the behaviour b;
- $\hat{n}_{b}^{\mathrm{EM}} = p(b_1 = b | \mathbf{w}_{1:J_{tr}}, \boldsymbol{\Omega}^{\mathrm{old}})$—the "expected number of times", when the behaviour b is associated to the first document, in this case the "expected number" is just a probability, the notation is used for the similarity with the rest of the parameters;
- $\hat{n}_{b'b}^{\mathrm{EM}} = \sum\limits_{j=2}^{J_{tr}} p(b_j = b', b_{j-1} = b | \mathbf{w}_{1:J_{tr}}, \boldsymbol{\Omega}^{\mathrm{old}})$—the expected number of times, when the behaviour b is followed by the behaviour b'.

During the E-step with the fixed current estimates of parameters $\boldsymbol{\Omega}^{\mathrm{old}}$, the updated values for the posterior distributions of the hidden variables should be computed. The derivation of the updated formulae for these distributions is similar to the Baum-Welch forward-backward algorithm [8], where the EM-algorithm is applied to get the MLE for a hidden Markov model. This similarity appears because the generative model can be viewed as an extension of a hidden Markov model.

For effective computation of the required posterior distributions the additional variables $\acute{\alpha}_b(j)$ and $\acute{\beta}_b(j)$ are introduced. A dynamic programming technique is applied for computation of these variables. Having the updated values for $\acute{\alpha}_b(j)$ and $\acute{\beta}_b(j)$ one can update the required posterior distributions of the hidden variables. The E-step is then formulated as follows (for simplification of notation the superscript "old" for the parameter variables is omitted inside the formulae):

$$\begin{cases} \acute{\alpha}_b(j) = \prod_{i=1}^{N_j} \sum_{k\in\mathcal{K}} \phi_{w_{ji}\,k}\,\theta_{kb} \sum_{b'\in\mathcal{B}} \acute{\alpha}_{b'}(j-1)\xi_{b\,b'}, \text{ if } j \geq 2,\ \forall b \in \mathcal{B}; \\ \acute{\alpha}_b(1) = \omega_b \prod_{i=1}^{N_1} \sum_{k\in\mathcal{K}} \phi_{w_{1i}\,k}\,\theta_{kb},\ \forall b \in \mathcal{B}; \end{cases} \tag{3.10}$$

$$\begin{cases} \acute{\beta}_b(j) = \sum_{b'\in\mathcal{B}} \acute{\beta}_{b'}(j+1)\xi_{b'\,b} \prod_{i=1}^{N_{j+1}} \sum_{k\in\mathcal{K}} \phi_{w_{j+1i}k}\,\theta_{kb'}, \text{ if } j \leq J-1,\ \forall b \in \mathcal{B}; \\ \acute{\beta}_b(J) = 1,\ \forall b \in \mathcal{B}; \end{cases} \tag{3.11}$$

$$Z = \sum_{b\in\mathcal{B}} \acute{\alpha}_b(1)\acute{\beta}_b(1); \tag{3.12}$$

$$p(b_1|\mathbf{w}_{1:J}, \mathbf{\Omega}^{\text{old}}) = \frac{\acute{\alpha}_{b_1}(1)\acute{\beta}_{b_1}(1)}{Z}; \tag{3.13}$$

$$p(b_j, b_{j-1}|\mathbf{w}_{1:J}, \mathbf{\Omega}^{\text{old}}) = \frac{\acute{\alpha}_{b_{j-1}}(j-1)\acute{\beta}_{b_j}(j)\xi_{b_j\,b_{j-1}}}{Z} \prod_{i=1}^{N_j} \sum_{k\in\mathcal{K}} \phi_{w_{ji}\,k}\theta_{kb_j}; \tag{3.14}$$

$$\begin{cases} p(z_{ji}, b_j|\mathbf{w}_{1:J_{tr}}, \mathbf{\Omega}^{\text{old}}) = \frac{1}{Z}\phi_{w_{ji}\,z_{ji}}\theta_{z_{ji}\,b_j}\acute{\beta}_{b_j}(j) \sum_{b'\in\mathcal{B}} \acute{\alpha}_{b'}(j-1)\xi_{b_j\,b'} \\ \quad \prod_{\substack{i'=1 \\ i'\neq i}}^{N_j} \sum_{k'\in\mathcal{K}} \phi_{w_{ji'}\,k'}\theta_{k'\,b_j}, \text{ if } j \geq 2; \\ p(z_{1i}, b_1|\mathbf{w}_{1:J_{tr}}, \mathbf{\Omega}^{\text{old}}) = \frac{1}{Z}\phi_{w_{1i}\,z_{1i}}\theta_{z_{1i}\,b_1}\acute{\beta}_{b_1}(1)\omega_{b_1} \prod_{\substack{i'=1 \\ i'\neq i}}^{N_1} \sum_{k'\in\mathcal{K}} \phi_{w_{1i'}\,k'}\theta_{k'\,b_1}; \end{cases} \tag{3.15}$$

$$p(z_{ji}|\mathbf{w}_{1:J}, \mathbf{\Omega}^{\text{old}}) = \sum_{b\in\mathcal{B}} p(z_{ji}, b|\mathbf{w}_{1:J}, \mathbf{\Omega}^{\text{old}}), \tag{3.16}$$

where Z is a normalisation constant for all the posterior distributions of the hidden variables. The details of the derivation are given in Appendix A.

Starting with some random initialisation of the parameter estimates, the EM-algorithm iterates the E and M-steps until convergence. The obtained estimates of parameters are used for further analysis.

3.3.2 Variational Inference

We also propose a learning algorithm based on the variational Bayes (VB) approach [9] to find approximated posterior distributions for both the hidden variables and the parameters.

In the VB inference scheme the true posterior distribution, in this case the distribution of the parameters and the hidden variables $p(\mathbf{z}_{1:J_{tr}}, \mathbf{b}_{1:J_{tr}}, \boldsymbol{\Omega}|\mathbf{w}_{1:J_{tr}}, \varkappa, \upsilon, \alpha, \eta)$, is approximated with a factorised distribution—$q(\mathbf{z}_{1:J_{tr}}, \mathbf{b}_{1:J_{tr}}, \boldsymbol{\Omega})$. The approximation is made to minimise the Kullback-Leibler divergence between the factorised and true distributions. We factorise the distribution in order to separate the hidden variables and the parameters:

$$\hat{q}(\mathbf{z}_{1:J_{tr}}, \mathbf{b}_{1:J_{tr}}, \boldsymbol{\Omega}) = \hat{q}(\mathbf{z}_{1:J_{tr}}, \mathbf{b}_{1:J_{tr}})\hat{q}(\boldsymbol{\Omega}) \stackrel{\text{def}}{=} \\ \operatorname{argmin} \mathcal{KL}\left(q(\mathbf{z}_{1:J_{tr}}, \mathbf{b}_{1:J_{tr}})q(\boldsymbol{\Omega})||p(\mathbf{z}_{1:J_{tr}}, \mathbf{b}_{1:J_{tr}}, \boldsymbol{\Omega}|\mathbf{w}_{1:J_{tr}}, \varkappa, \upsilon, \alpha, \eta)\right), \tag{3.17}$$

where $\mathcal{KL}$ denotes the Kullback-Leibler divergence. The minimisation of the Kullback-Leibler divergence is equivalent to the maximisation of the evidence lower bound (ELBO). The maximisation is done by coordinate ascent [9].

During the update of the parameters the approximated distribution $q(\boldsymbol{\Omega})$ is further factorised:

$$q(\boldsymbol{\Omega}) = q(\omega)q(\Xi)q(\boldsymbol{\Theta})q(\boldsymbol{\Phi}). \tag{3.18}$$

Note that this factorisation is a corollary of our model and not an assumption.

The iterative process of updating the approximated distributions of the parameters and the hidden variables can be formulated as an EM-like algorithm, where during the E-step the approximated distributions of the hidden variables are updated and during the M-step the approximated distributions of the parameters are updated.

The M-like step is as follows:

$$\begin{cases} q(\boldsymbol{\Phi}) = \prod_{k \in \mathcal{K}} Dir\left(\phi_k; \tilde{\eta}_k\right), \\ \tilde{\eta}_{wk} = \eta_w + \hat{l}_{wk}^{\text{VB}}, \qquad \forall w \in \mathcal{V}, k \in \mathcal{K}; \end{cases} \tag{3.19}$$

$$\begin{cases} q(\boldsymbol{\Theta}) = \prod_{b \in \mathcal{B}} Dir(\theta_b; \tilde{\alpha}_b), \\ \tilde{\alpha}_{kb} = \alpha_k + \hat{n}_{kb}^{\text{VB}}, \qquad \forall k \in \mathcal{K}, b \in \mathcal{B}; \end{cases} \tag{3.20}$$

$$\begin{cases} q(\omega) = Dir(\omega; \tilde{\varkappa}), \\ \tilde{\varkappa}_b = \varkappa_b + \hat{n}_b^{\text{VB}}, \qquad \forall b \in \mathcal{B}; \end{cases} \tag{3.21}$$

$$\begin{cases} q(\Xi) = \prod_{b \in \mathcal{B}} Dir(\xi_b; \tilde{\upsilon}_b), \\ \tilde{\upsilon}_{b'b} = \upsilon_{b'} + \hat{n}_{b'b}^{\text{VB}}, \qquad \forall b', b \in \mathcal{B}, \end{cases} \tag{3.22}$$

where $\tilde{\boldsymbol{\eta}}_z$, $\tilde{\boldsymbol{\alpha}}_b$, $\tilde{\boldsymbol{\varkappa}}$ and $\tilde{\boldsymbol{\upsilon}}_b$ are updated hyperparameters of the corresponding posterior Dirichlet distributions and

- $\hat{l}_{wk}^{\text{VB}} = \sum_{j=1}^{J_{tr}} \sum_{i=1}^{N_j} \mathbb{1}(w_{ji} = w) q(z_{ji} = k)$—the expected number of times, when the word w is associated with the topic k. Here and below the expected number is computed with respect to the approximated posterior distributions of the hidden variables;
- $\hat{n}_{kb}^{\text{VB}} = \sum_{j=1}^{J_{tr}} \sum_{i=1}^{N_j} q(z_{ji} = k, b_j = b)$—the expected number of times, when the topic k is associated with the behaviour b;
- $\hat{n}_{b}^{\text{VB}} = q(b_1 = b)$—the "expected number" of times, when the behaviour b is associated with the first document;
- $\hat{n}_{b'b}^{\text{VB}} = \sum_{j=2}^{J_{tr}} q(b_j = b', b_{j-1} = b)$—the expected number of times, when the behaviour b is followed by the behaviour b'.

The following additional variables are introduced for the E-like step:

$$\tilde{\omega}_b = \exp\left(\psi\left(\tilde{\varkappa}_b\right) - \psi\left(\sum_{b' \in \mathcal{B}} \tilde{\varkappa}_{b'}\right)\right), \qquad \forall b \in \mathcal{B}; \qquad (3.23)$$

$$\tilde{\xi}_{\tilde{b}b} = \exp\left(\psi\left(\tilde{\upsilon}_{\tilde{b}b}\right) - \psi\left(\sum_{b' \in \mathcal{B}} \tilde{\upsilon}_{b'b}\right)\right), \qquad \forall \tilde{b}, b \in \mathcal{B}; \qquad (3.24)$$

$$\tilde{\phi}_{wk} = \exp\left(\psi\left(\tilde{\eta}_{wk}\right) - \psi\left(\sum_{w' \in \mathcal{V}} \tilde{\eta}_{w'k}\right)\right), \qquad \forall w \in \mathcal{V}, k \in \mathcal{K}; \qquad (3.25)$$

$$\tilde{\theta}_{kb} = \exp\left(\psi\left(\tilde{\alpha}_{kb}\right) - \psi\left(\sum_{k' \in \mathcal{K}} \tilde{\alpha}_{k'b}\right)\right), \qquad \forall k \in \mathcal{K}, b \in \mathcal{B}. \qquad (3.26)$$

where $\psi(\cdot)$ is the digamma function.

Using these additional notations, the E-like step is formulated in the same way as the E-step of the EM-algorithm, replacing everywhere the estimates of parameters with the corresponding tilde introduced notation and true posterior distributions of the hidden variables with the corresponding approximated ones in (3.10)–(3.16). The full details of the VB-algorithm derivation are presented in Appendix B.

The point estimates of parameters can be obtained by expected values of the posterior approximated distributions. An expected value for a Dirichlet distribution (a posterior distribution for all the parameters) is a normalised vector of hyperparameters. Using the expressions for the hyperparameters from (3.19)–(3.22), thc final parameter estimates can be obtained by:

$$\widehat{\phi}_{wk}^{\text{VB}} = \frac{\eta_w + \hat{l}_{wk}^{\text{VB}}}{\sum\limits_{w' \in \mathcal{V}} \left(\eta_{w'} + \hat{l}_{w'k}^{\text{VB}}\right)}, \qquad \forall w \in \mathcal{V}, k \in \mathcal{K}; \tag{3.27}$$

$$\widehat{\theta}_{kb}^{\text{VB}} = \frac{\alpha_k + \hat{n}_{kb}^{\text{VB}}}{\sum\limits_{k' \in \mathcal{K}} \left(\alpha_{k'} + \hat{n}_{k'b}^{\text{VB}}\right)}, \qquad \forall k \in \mathcal{K}, b \in \mathcal{B}; \tag{3.28}$$

$$\widehat{\xi}_{b'b}^{\text{VB}} = \frac{\upsilon_{b'} + \hat{n}_{b'b}^{\text{VB}}}{\sum\limits_{\tilde{b} \in \mathcal{B}} \left(\upsilon_{\tilde{b}} + \hat{n}_{\tilde{b}b}^{\text{VB}}\right)}, \qquad \forall b', b \in \mathcal{B}; \tag{3.29}$$

$$\widehat{\omega}_{b}^{\text{VB}} = \frac{\varkappa_b + \hat{n}_{b}^{\text{VB}}}{\sum\limits_{b' \in \mathcal{B}} \left(\varkappa_{b'} + \hat{n}_{b'}^{\text{VB}}\right)}, \qquad \forall b \in \mathcal{B}. \tag{3.30}$$

3.3.3 Gibbs Sampling

In [2] the collapsed version of Gibbs sampling (GS) is used for parameter learning in the MCTM. The Markov chain is built to sample only the hidden variables z_{ji} and b_j, while the parameters $\mathbf{\Phi}$, $\mathbf{\Theta}$ and Ξ are integrated out (note that the distribution for the initial behaviour choice ω is not considered in [2]).

During the burn-in stage the hidden topic and behaviour assignments to each token in the dataset are drawn from the conditional distributions given all the remaining variables. Following the Markov Chain Monte Carlo framework it would draw samples from the posterior distribution $p(\mathbf{z}_{1:J_{tr}}, \mathbf{b}_{1:J_{tr}} | \mathbf{w}_{1:J_{tr}}, \boldsymbol{\eta}, \boldsymbol{\alpha}, \boldsymbol{\varkappa}, \boldsymbol{\upsilon})$. From the whole sample for $\{\mathbf{z}_{1:J_{tr}}, \mathbf{b}_{1:J_{tr}}\}$ the parameters can be estimated by [10]:

$$\widehat{\phi}_{wk}^{\text{GS}} = \frac{\hat{l}_{wk}^{\text{GS}} + \eta_w}{\sum\limits_{w' \in \mathcal{V}} \left(\hat{n}_{w'k}^{\text{GS}} + \eta_{w'}\right)}, \qquad \forall w \in \mathcal{V}, k \in \mathcal{K}; \tag{3.31}$$

$$\widehat{\theta}_{kb}^{\text{GS}} = \frac{\hat{n}_{kb}^{\text{GS}} + \alpha_k}{\sum\limits_{k' \in \mathcal{K}} \left(\hat{n}_{k'b}^{\text{GS}} + \alpha_{k'}\right)}, \qquad \forall k \in \mathcal{K}, b \in \mathcal{B}; \tag{3.32}$$

$$\widehat{\xi}_{b'b}^{\text{GS}} = \frac{\hat{n}_{b'b}^{\text{GS}} + \upsilon_{b'}}{\sum\limits_{\tilde{b} \in \mathcal{B}} \left(\hat{n}_{\tilde{b}b}^{\text{GS}} + \upsilon_{\tilde{b}}\right)}, \qquad \forall b', b \in \mathcal{B}, \tag{3.33}$$

where $\hat{l}_{wk}^{\text{GS}}$ is the count for the number of times when the word w is associated with the topic k, $\hat{n}_{kb}^{\text{GS}}$ is the count for the topic k and the behaviour b pair, $\hat{n}_{b'b}^{\text{GS}}$ is the count for the number of times when the behaviour b is followed by the behaviour b'.

The Rao-Blackwell-Kolmogorov theorem guarantees that the variance of the estimates of parameters obtained by a collapsed Gibbs sampler is never higher that the variance of a sample of the parameters from a full Gibbs sampler [11] and these estimates can be treated as posterior samples.

3.3.4 Similarities and Differences of the Learning Algorithms

The point parameter estimates for all three learning algorithms (3.6)–(3.9), (3.27)–(3.30) and (3.31)–(3.33) have a similar form. The EM-algorithm estimates differ up to the hyperparameter reassignment—adding one to all the hyperparameters in the VB or GS algorithms ends up with the same final equations for the parameter estimates in the EM-algorithm. We explore this in the experimental part. This "-1" term in the EM-algorithm formulae (3.6)–(3.8) occurs because it uses modes of the posterior distributions while the point estimates obtained by the VB and GS algorithms are means of the corresponding posterior distributions. For a Dirichlet distribution, which is a posterior distribution for all the parameters, mode and mean expressions differ in this "-1" term.

The main differences of the methods consist in the ways the counts l_{wk}, n_{kb} and $n_{b'b}$ are estimated. In the GS algorithm they are calculated by a single sample from the posterior distribution of the hidden variables $p(\mathbf{z}_{1:J_{tr}}, \mathbf{b}_{1:J_{tr}}|\mathbf{w}_{1:J_{tr}}, \boldsymbol{\eta}, \boldsymbol{\alpha}, \boldsymbol{\upsilon})$. In the EM-algorithm the counts are computed as expected numbers of the corresponding events with respect to the posterior distributions of the hidden variables. In the VB algorithm the counts are computed in the same way as in the EM-algorithm up to replacing the true posterior distributions with the approximated ones.

Our observations for the dynamic topic model confirm the comparison results for the vanilla PLSA and LDA models provided in [12].

3.4 Anomaly Detection

Online anomaly detection can be performed with the MCTM. Here anomalies in video streams are considered. The decision making procedure is divided into two stages. In the learning stage the parameters are estimated using J_{tr} visual documents by one of the learning algorithms, presented in Sect. 3.3. After that during the testing stage a decision about abnormality of new upcoming test documents is made, comparing the predictive likelihood of each document with a threshold. The likelihood is computed using the parameters obtained during the learning stage. The threshold is a parameter of the method and can be set empirically, for example, to label 2% of the test data as abnormal. A comparison of the algorithms based on a measure independent of threshold value selection is presented in Sect. 3.5.

We also propose an anomaly localisation procedure during the testing stage for those visual documents that are labelled as abnormal. This procedure is designed to provide spatial information about anomalies, while documents labelled as abnormal provide temporal detection. The following sections introduce both the anomaly detection procedure at a document level and the anomaly localisation procedure within a video frame.

3.4.1 Abnormal Documents Detection

The predictive likelihood of a new visual document $\mathbf{w}_j$ given all the previous data $\mathbf{w}_{1:j-1}$ can be used as normality measure of the document [2]:

$$p(\mathbf{w}_j|\mathbf{w}_{1:j-1}) = \iiint p(\mathbf{w}_j|\mathbf{w}_{1:j-1}, \boldsymbol{\Phi}, \boldsymbol{\Theta}, \boldsymbol{\Xi})p(\boldsymbol{\Phi}, \boldsymbol{\Theta}, \boldsymbol{\Xi}|\mathbf{w}_{1:j-1})\mathrm{d}\boldsymbol{\Phi}\mathrm{d}\boldsymbol{\Theta}\mathrm{d}\boldsymbol{\Xi}. \tag{3.34}$$

If the likelihood value is small it means that the current document cannot be fitted to the learnt behaviours and topics, which represent typical motion patterns. Therefore, this is an indication for an abnormal event in this document. The decision about abnormality of a document is then made by comparing the predictive likelihood of the document with the threshold.

In real world applications it is essential to detect anomalies as soon as possible. Hence an approximation of the integral in (3.34) is used for efficient computation. The first approximation is based on the assumption that the training dataset is representative for parameter learning, which means that posterior probability of the parameters would not change if there is more observed data:

$$p(\boldsymbol{\Phi}, \boldsymbol{\Theta}, \boldsymbol{\Xi}|\mathbf{w}_{1:j-1}) \approx p(\boldsymbol{\Phi}, \boldsymbol{\Theta}, \boldsymbol{\Xi}|\mathbf{w}_{1:J_{tr}}) \quad \forall j > J_{tr}. \tag{3.35}$$

The predictive likelihood can be then approximated as:

$$\iiint p(\mathbf{w}_j|\mathbf{w}_{1:j-1}, \boldsymbol{\Phi}, \boldsymbol{\Theta}, \boldsymbol{\Xi})p(\boldsymbol{\Phi}, \boldsymbol{\Theta}, \boldsymbol{\Xi}|\mathbf{w}_{1:j-1})\mathrm{d}\boldsymbol{\Phi}\mathrm{d}\boldsymbol{\Theta}\mathrm{d}\boldsymbol{\Xi} \approx \iiint p(\mathbf{w}_j|\mathbf{w}_{1:j-1}, \boldsymbol{\Phi}, \boldsymbol{\Theta}, \boldsymbol{\Xi})p(\boldsymbol{\Phi}, \boldsymbol{\Theta}, \boldsymbol{\Xi}|\mathbf{w}_{1:J_{tr}})\mathrm{d}\boldsymbol{\Phi}\mathrm{d}\boldsymbol{\Theta}\mathrm{d}\boldsymbol{\Xi}. \tag{3.36}$$

Depending on the algorithm used for learning the integral in (3.36) can be further approximated in different ways. We consider two types of approximation.

3.4.1.1 Plug-In Approximation

The point estimates of parameters can be plugged into in the integral (3.36) for approximation:

$$\iiint p(\mathbf{w}_j|\mathbf{w}_{1:j-1}, \boldsymbol{\Phi}, \boldsymbol{\Theta}, \boldsymbol{\Xi})p(\boldsymbol{\Phi}, \boldsymbol{\Theta}, \boldsymbol{\Xi}|\mathbf{w}_{1:J_{tr}})\mathrm{d}\boldsymbol{\Phi}\mathrm{d}\boldsymbol{\Theta}\mathrm{d}\boldsymbol{\Xi} \approx$$
$$\iiint p(\mathbf{w}_j|\mathbf{w}_{1:j-1}, \boldsymbol{\Phi}, \boldsymbol{\Theta}, \boldsymbol{\Xi})\delta_{\hat{\boldsymbol{\Phi}}}(\boldsymbol{\Phi})\delta_{\hat{\boldsymbol{\Theta}}}(\boldsymbol{\Theta}), \delta_{\hat{\boldsymbol{\Xi}}}(\boldsymbol{\Xi})\mathrm{d}\boldsymbol{\Phi}\mathrm{d}\boldsymbol{\Theta}\mathrm{d}\boldsymbol{\Xi} = p(\mathbf{w}_j|\mathbf{w}_{1:j-1}, \hat{\boldsymbol{\Phi}}, \hat{\boldsymbol{\Theta}}, \hat{\boldsymbol{\Xi}}), \tag{3.37}$$

where $\delta_a(\cdot)$ is the delta-function with the centre in a; $\hat{\boldsymbol{\Phi}}$, $\hat{\boldsymbol{\Theta}}$ and $\hat{\boldsymbol{\Xi}}$ are point estimates of parameters, which can be computed by any of the considered learning algorithms using (3.6)–(3.8), (3.27)–(3.29) or (3.31)–(3.33).

The product and sum rules, the conditional independence equations from the generative model are then applied and the final formula for the plug-in approximation is follows:

$$p(\mathbf{w}_j|\mathbf{w}_{1:j-1}) \approx p(\mathbf{w}_j|\mathbf{w}_{1:j-1}, \hat{\boldsymbol{\Phi}}, \hat{\boldsymbol{\Theta}}, \hat{\boldsymbol{\Xi}}) =$$
$$\sum_{b_{j-1}}\sum_{b_j}\Big[p(\mathbf{w}_j|b_j, \hat{\boldsymbol{\Phi}}, \hat{\boldsymbol{\Theta}})p(b_j|b_{j-1}, \hat{\boldsymbol{\Xi}})p(b_{j-1}|\mathbf{w}_{1:j-1}, \hat{\boldsymbol{\Phi}}, \hat{\boldsymbol{\Theta}}, \hat{\boldsymbol{\Xi}})\Big], \tag{3.38}$$

where the predictive probability of the behaviour for the current document, given the observed data up to the current document, can be computed via the recursive formula:

$$p(b_{j-1}|\mathbf{w}_{1:j-1}, \hat{\boldsymbol{\Phi}}, \hat{\boldsymbol{\Theta}}, \hat{\boldsymbol{\Xi}}) =$$
$$\sum_{b_{j-2}}\frac{p(\mathbf{w}_{j-1}|b_{j-1}, \hat{\boldsymbol{\Phi}}, \hat{\boldsymbol{\Theta}})p(b_{j-1}|b_{j-2}, \hat{\boldsymbol{\Xi}})p(b_{j-2}|\mathbf{w}_{1:j-2}, \hat{\boldsymbol{\Phi}}, \hat{\boldsymbol{\Theta}}, \hat{\boldsymbol{\Xi}})}{p(\mathbf{w}_{j-1}|\mathbf{w}_{1:j-2}, \hat{\boldsymbol{\Phi}}, \hat{\boldsymbol{\Theta}}, \hat{\boldsymbol{\Xi}})}. \tag{3.39}$$

The point estimates can be computed for all three learning algorithms, therefore a normality measure based on the plug-in approximation of the predictive likelihood is applicable for all of them.

3.4.1.2 Monte Carlo Approximation

If samples $\{\boldsymbol{\Phi}^s, \boldsymbol{\Theta}^s, \boldsymbol{\Xi}^s\}_{s=1}^S$, where S is the number of samples, from the posterior distribution $p(\boldsymbol{\Phi}, \boldsymbol{\Theta}, \boldsymbol{\Xi}|\mathbf{w}_{1:J_{tr}})$ of the parameters can be obtained, the integral (3.36) can be further approximated by the Monte Carlo method:

$$\iiint p(\mathbf{w}_j|\mathbf{w}_{1:j-1}, \boldsymbol{\Phi}, \boldsymbol{\Theta}, \boldsymbol{\Xi})p(\boldsymbol{\Phi}, \boldsymbol{\Theta}, \boldsymbol{\Xi}|\mathbf{w}_{1:J_{tr}})\mathrm{d}\boldsymbol{\Phi}\mathrm{d}\boldsymbol{\Theta}\mathrm{d}\boldsymbol{\Xi} \approx$$
$$\frac{1}{S}\sum_{s=1}^{S} p(\mathbf{w}_j|\mathbf{w}_{1:j-1}, \boldsymbol{\Phi}^s, \boldsymbol{\Theta}^s, \boldsymbol{\Xi}^s). \tag{3.40}$$

These samples can be obtained (a) from the approximated posterior distributions $q(\boldsymbol{\Phi})$, $q(\boldsymbol{\Theta})$, and $q(\boldsymbol{\Xi})$ of the parameters, computed by the VB learning algorithm, or (b) from the independent samples of the GS scheme. For the conditional likelihood $p(\mathbf{w}_j|\mathbf{w}_{1:j-1}, \boldsymbol{\Phi}^s, \boldsymbol{\Theta}^s, \boldsymbol{\Xi}^s)$ the formula (3.38) is valid.

Note that for the approximated posterior distribution of the parameters, i.e., the output of the VB learning algorithm, the integral (3.36) can be resolved analytically, but it would be computationally infeasible. This is the reason why the Monte Carlo approximation is used in this case.

Finally, in order to compare documents of different lengths the normalised likelihood is used as a normality measure $\mathcal{A}$:

$$\mathcal{A}(\mathbf{w}_j) = \frac{1}{N_j} \log p(\mathbf{w}_j|\mathbf{w}_{1:j-1}). \tag{3.41}$$

3.4.2 *Localisation of Anomalies*

The topic modeling approach allows to compute a likelihood function not only for the whole document but for an individual word within the document too. Recall that the visual word contains the information about a location in the frame. We propose to use the location information from the least probable words (e.g., 10 words with the lowest likelihood values) to localise anomalies in the frame. Note that we do not require anything additional to a topic model, e.g., modelling regional information explicitly as in [13] or comparing a test document with training ones as in [14]. Instead, the proposed anomaly localisation procedure is general and can be applied in any topic modeling based method, where spatial information is encoded as visual words.

The predictive likelihood of a word can be computed in a similar way to the likelihood of the whole document. For the point estimates of parameters and plug-in approximation of the integral it is:

$$p(w_{ji}|\mathbf{w}_{1:j-1}) \approx p(w_{ji}|\mathbf{w}_{1:j-1}, \hat{\boldsymbol{\Phi}}, \hat{\boldsymbol{\Theta}}, \hat{\boldsymbol{\Xi}}). \tag{3.42}$$

For the samples from the posterior distributions of the parameters and the Monte Carlo integral approximation it is:

$$p(w_{ji}|\mathbf{w}_{1:j-1}) \approx \frac{1}{S} \sum_{s=1}^{S} p(w_{ji}|\mathbf{w}_{1:j-1}, \boldsymbol{\Phi}^s, \boldsymbol{\Theta}^s, \boldsymbol{\Xi}^s). \tag{3.43}$$

3.5 Performance Validation

We compare the two proposed learning algorithms, based on EM and VB, with the GS algorithm, proposed in [2], on two real datasets.

The performance of the algorithms is compared on the QMUL street intersection data [2] and Idiap traffic junction data [5]. Both datasets are 45-minutes video sequences, captured of busy traffic road junctions, where we use a 5-minute video sequence as a training dataset and others as a test. The documents that have less than 20 visual words are discarded from consideration. In practice these documents can be classified to be normal by default as there is not enough information to make a decision. The frame size for both datasets is 288×360. Sample frames are presented in Fig. 3.3.

The size of grid cells is set to 8×8 pixels for spatial quantisation of the local motion for visual word determination. Non-overlapping clips with a one second length are treated as visual documents.

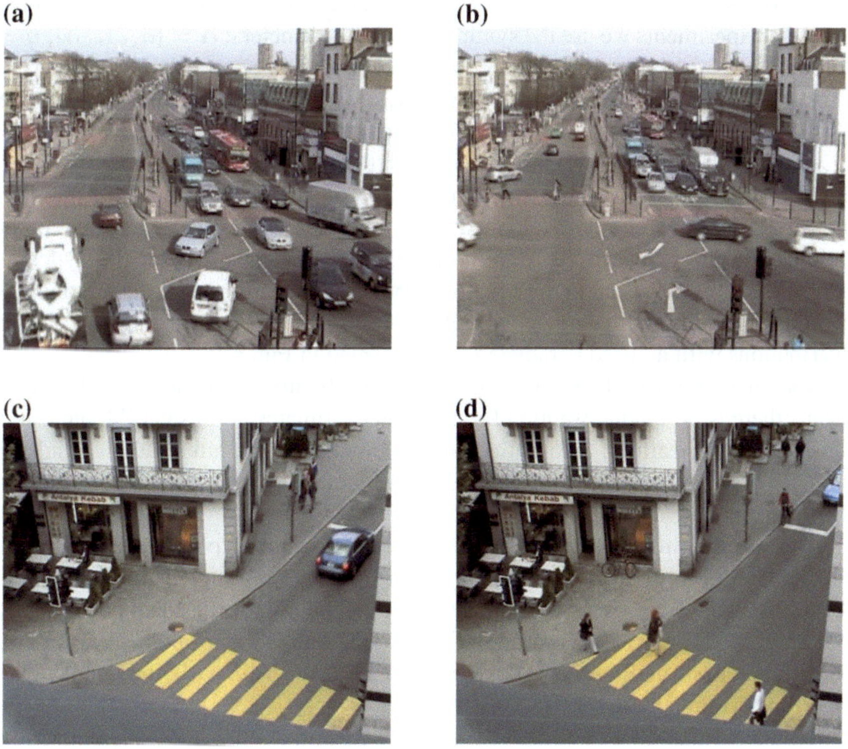

Fig. 3.3 Sample frames of the real datasets. The top row presents two sample frames from the QMUL data, the bottom row presents two sample frames from the Idiap data

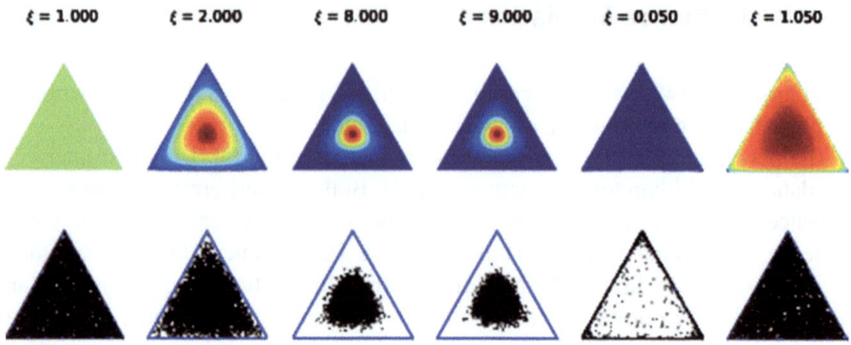

Fig. 3.4 Dirichlet distributions with different symmetric parameters ξ. For the representation purposes the three-dimensional space is used. On the top row the colours correspond to the Dirichlet probability density function values in the area. On the bottom row there are samples generated from the corresponding density functions. The sample size is 5000

We also study the influence of the hyperparameters on the learning algorithms. In all the experiments we use the symmetric hyperparameters: $\boldsymbol{\alpha} = \{\alpha, \dots, \alpha\}$, $\boldsymbol{\eta} = \{\eta, \dots, \eta\}$, $\boldsymbol{\upsilon} = \{\upsilon, \dots, \upsilon\}$ and $\varkappa = \{\varkappa, \dots, \varkappa\}$. The three groups of the hyperparameter settings are compared: $\{\alpha = 1, \eta = 1, \upsilon = 1, \varkappa = 1\}$ (referred as "prior type 1"), $\{\alpha = 8, \eta = 0.05, \upsilon = 1, \varkappa = 1\}$ ("prior type H") and $\{\alpha = 9, \eta = 1.05, \upsilon = 2, \varkappa = 2\}$ ("prior type H+1"). Note that the first group corresponds to the case when in the EM-algorithm learning scheme the prior components are cancelled out, i.e., the MAP estimates in this case are equal to the MLE. The equations for the point estimates in the EM learning algorithm with the prior type H+1 of the hyperparameter settings are equal to the equations for the point estimates in the VB and GS learning algorithms with the prior type H of the settings. The corresponding Dirichlet distributions with all used parameters are presented in Fig. 3.4.

Note that parameter learning is an ill-posed problem in topic modeling [7]. This means there is no unique solution for parameter estimates. We use 20 Monte Carlo runs for all the learning algorithms with different random initialisations resulting with different solutions. The mean results among these runs are presented below for comparison.

All three algorithms are run with three different groups of hyperparameter settings. The number of topics and behaviours is set to 8 and 4, respectively, for the QMUL dataset, 10 and 3 are used for the corresponding values for the Idiap dataset. The EM and VB algorithms are run for 100 iterations. The GS algorithm is run for 500 burn-in iterations and 5 independent samples are taken with a 100 iterations delay after the burn-in period.

3.5.1 Performance Measure

Anomaly detection performance of the algorithms depends on threshold selection. To make a fair comparison of the different learning algorithms we use a performance measure which is independent of threshold selection.

In binary classification the following measures [15] are used:

- TP—true positive, the number of documents, which are correctly detected as positive (abnormal in our case);
- TN—true negative, the number of documents, which are correctly detected as negative (normal in our case);
- FP—false positive, the number of documents, which are incorrectly detected as positive, when they are negative;
- FN—false negative, the number of documents, which are incorrectly detected as negative, when they are positive;
- $\text{precision} = \frac{\text{TP}}{\text{TP} + \text{FP}}$—a fraction of correct detections among all documents labelled as abnormal by an algorithm;
- $\text{recall} = \frac{\text{TP}}{\text{TP} + \text{FN}}$—a fraction of correct detections among all truly abnormal documents.

The area under the precision-recall curve is used as a performance measure. This measure is more informative for detection of rare events than the popular area under the receiver operating characteristic (ROC) curve [15].

3.5.2 Parameter Learning

We visualise the learnt behaviours for the qualitative assessment of the proposed framework (Figs. 3.5 and 3.6). For illustrative purposes we consider one run of the EM learning algorithm with the prior type H+1 of the hyperparameter settings.

The behaviours learnt on the QMUL data are shown in Fig. 3.5 (for visualisation words representing 50% of a probability mass of a behaviour are used). One can notice that the algorithm correctly recognises the motion patterns in the data. The general motion of the scene follows a cycle: a vertical traffic flow (the first behaviour in Fig. 3.5a), when cars move downward and upward on the road; left and right turns (the fourth behaviour in Fig. 3.5d): some cars moving on the "vertical" road turn to the perpendicular road at the end of the vertical traffic flow; a left traffic flow (the second behaviour in Fig. 3.5b), when cars move from right to left on the "horizontal" road; and a right traffic flow (the third behaviour in Fig. 3.5c), when cars move from left to right on the "horizontal" road. Note that the ordering numbers of behaviours correspond to their internal representation in the algorithm. The transition probability matrix Ξ is used to recognise the correct order of behaviours in the data.

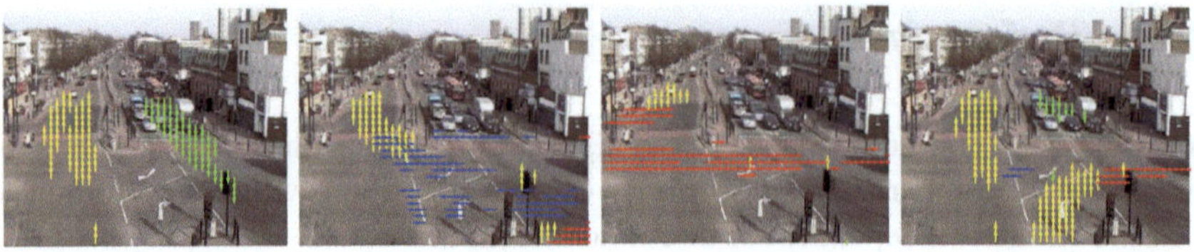

(a) Behaviour 1 (b) Behaviour 2 (c) Behaviour 3 (d) Behaviour 4

Fig. 3.5 Behaviours learnt by the EM learning algorithm on the QMUL data. The arrows represent the visual words: the location and direction of the motion. The first behaviour (**a**) corresponds to the vertical traffic flow, the second (**b**) and the third (**c**) behaviours correspond to the left and right traffic flow, respectively. The fourth (**d**) behaviour correspond to turns that follow the vertical traffic flow

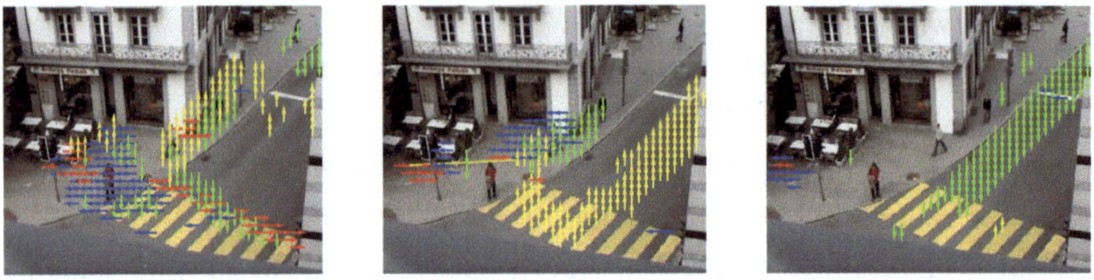

(a) Behaviour 1 (b) Behaviour 2 (c) Behaviour 3

Fig. 3.6 Behaviours learnt by the EM learning algorithm on the Idiap data. The arrows represent the visual words: the location and direction of the motion. The first behaviour (**a**) corresponds to a pedestrian motion, the second (**b**) and the third (**c**) behaviours correspond to the upward and downward traffic flows, respectively

Figure 3.6 presents the behaviours learnt on the Idiap data. In this case the learnt behaviours have also clear semantic meaning. The scene motion follows a cycle: a pedestrian flow (the first behaviour in Fig. 3.6a), when cars stop in front of the stop line and pedestrians cross the road; a downward traffic flow (the third behaviour in Fig. 3.6c), when cars move downward along the road; an upward traffic flow (the second behaviour in Fig. 3.6b), when cars from left and right sides move upward on the road.

3.5.3 Anomaly Detection

In this section the anomaly detection performance achieved by all three learning algorithms is compared. The datasets contain the number of abnormal events, such as jaywalking, car moving on the opposite lane and disruption of the traffic flow (see examples in Fig. 3.7).

For the EM learning algorithm the plug-in approximation of the predictive likelihood is used for anomaly detection. For both the VB and GS learning algorithms both the plug-in and Monte Carlo approximations of the likelihood are used. Note

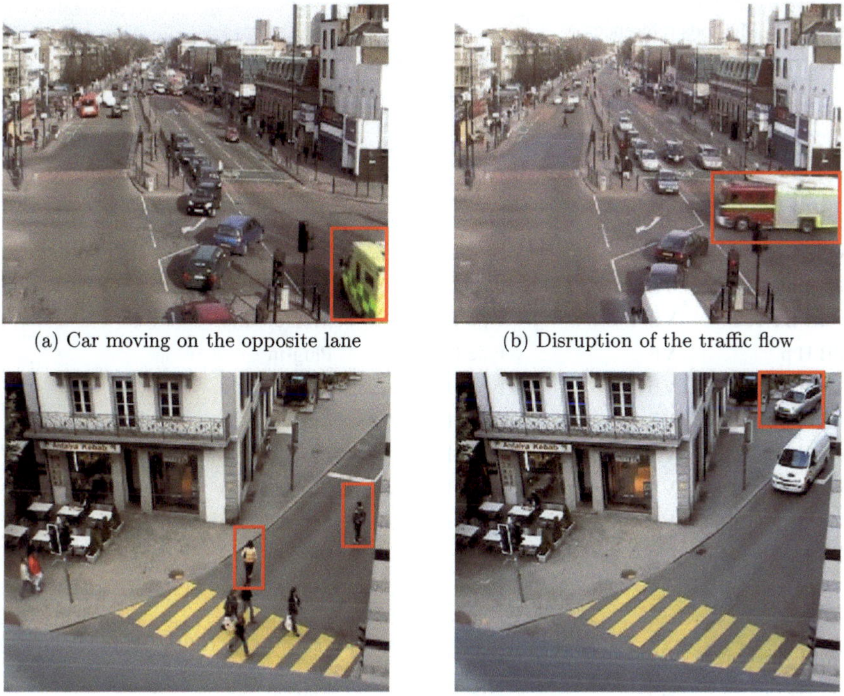

(a) Car moving on the opposite lane

(b) Disruption of the traffic flow

(c) Jaywalking

(d) Car moving on the sidewalk

Fig. 3.7 Examples of abnormal events

that for the GS algorithm samples are obtained during the learning stage. As 5 independent samples from the GS scheme are taken, the Monte Carlo approximation of the predictive likelihood is computed based on these 5 samples. For the VB learning algorithm samples are obtained after the learning stage from the posterior distributions, parameters of which are learnt. This means that the number of samples that are used for anomaly detection does not influence the computational cost of learning. We test the Monte Carlo approximation of the predictive likelihood with 5 and 100 samples for the VB learning algorithm.

As a result, we have 18 methods to compare: obtained by three learning algorithms, three different groups of hyperparameter settings, one type of predictive likelihood approximation for the EM learning algorithm, two types of predictive likelihood approximation for the VB and GS learning algorithms, where for the former there are two Monte Carlo approximations using 5 and 100 samples. The list of method references can be found in Table 3.1.

Note that we achieve a very fast decision making performance in our framework. Indeed, anomaly detection is made for approximately 0.0044 sec per visual document by the plug-in approximation of the predictive likelihood, for 0.0177 sec per

Table 3.1 Methods references

Reference	Learning algorithm	Hyperparameter settings	Marginal likelihood approximation	Number of posterior samples
EM 1 p	EM	type 1	Plug-in	–
EM H p	EM	type H	Plug-in	–
EM H+1 p	EM	type H+1	Plug-in	–
VB 1 p	VB	type 1	Plug-in	–
VB 1 mc 5	VB	type 1	Monte Carlo	5
VB 1 mc 100	VB	type 1	Monte Carlo	100
VB H p	VB	type H	Plug-in	–
VB H mc 5	VB	type H	Monte Carlo	5
VB H mc 100	VB	type H	Monte Carlo	100
VB H+1 p	VB	type H+1	Plug-in	–
VB H+1 mc 5	VB	type H+1	Monte Carlo	5
VB H+1 mc 100	VB	type H+1	Monte Carlo	100
GS 1 p	GS	type 1	Plug-in	–
GS 1 mc	GS	type 1	Monte Carlo	5
GS H p	GS	type H	Plug-in	–
GS H mc	GS	type H	Monte Carlo	5
GS H+1 p	GS	type H+1	Plug-in	–
GS H+1 mc	GS	type H+1	Monte Carlo	5

document by the Monte Carlo approximation with 5 samples and for 0.3331 sec per document by the Monte Carlo approximation with 100 samples.[2]

The mean areas under precision-recall curves for anomaly detection for all 18 compared methods can be found in Fig. 3.8. Below we examine the results with respect to hyperparameter sensitivity, an influence of the likelihood approximation on the final performance. We also compare the learning algorithms and discuss anomaly localisation results.

3.5.3.1 Hyperparameter Sensitivity

This section presents a sensitivity analysis of the anomaly detection methods with respect to changes of the hyperparameters.

The analysis of the mean areas under curves (Fig. 3.8) suggests that the hyperparameters almost do not influence the results of the EM learning algorithm, while there is significant dependence between hyperparameter changes and results of the

[2]The computational time is provided for a laptop computer with i7-4702HQ CPU with 2.20 GHz, 16 GB RAM using Matlab R2015a implementation.

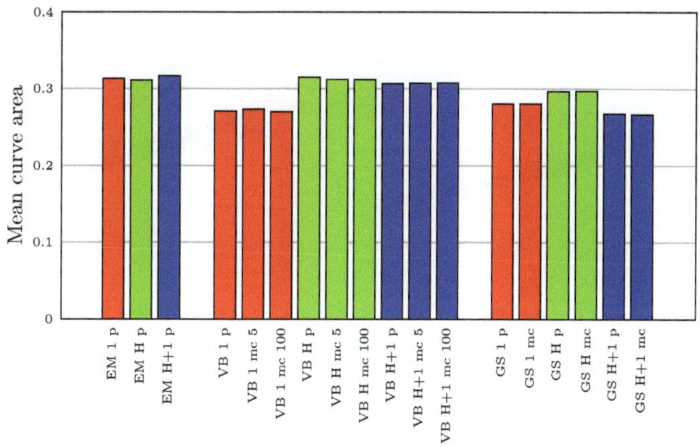

(a) QMUL data results

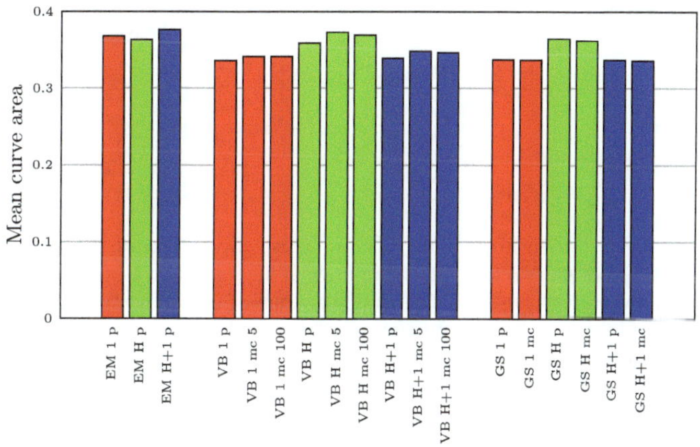

(b) Idiap data results

Fig. 3.8 Results of anomaly detection. The mean areas under precision-recall curves **a** on the QMUL data and **b** on the Idiap data

VB and GS learning algorithms. These conclusions are confirmed by examination of the individual runs of the algorithms. For example, Fig. 3.9 presents the precision-recall curves for all 20 runs with different initialisations of 4 methods on the Idiap data: the VB learning algorithm using the plug-in approximation of the predictive likelihood with the prior types 1 and H of the hyperparameter settings and the EM learning algorithm with the same prior groups of the hyperparameter settings. One can notice that the variance of the curves for the VB learning algorithm with the prior type 1 is larger than the corresponding variance with the prior type H, while the corresponding variances for the EM learning algorithm are very close to each other.

Note that the results of the EM learning algorithm with the prior type 1 do not significantly differ from the results with the other priors, despite of the fact that the prior type 1 actually cancels out the prior influence on the parameter estimates and

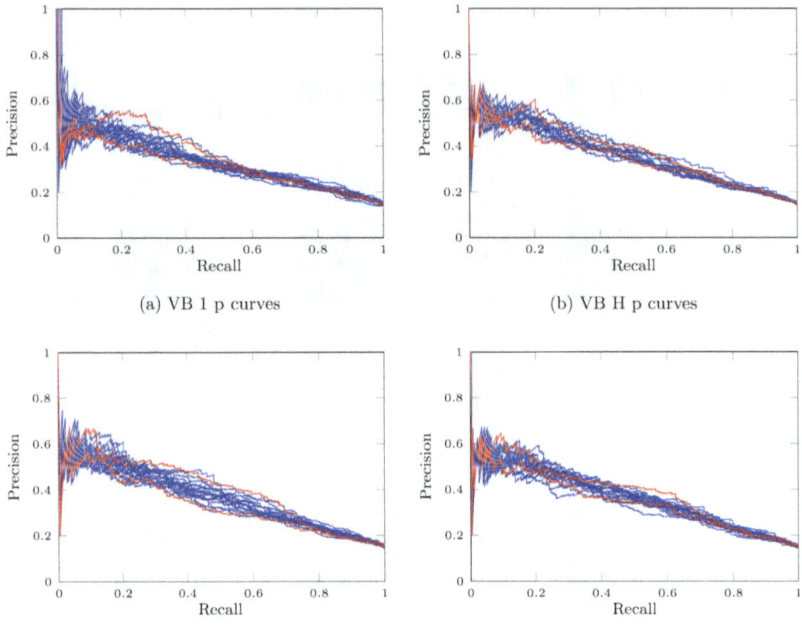

(a) VB 1 p curves

(b) VB H p curves

(c) EM 1 p curves

(d) EM H p curves

Fig. 3.9 Hyperparameter sensitivity of the precision-recall curves. The top row corresponds to all the independent runs of the VB learning algorithm with the prior type 1 (**a**) and the prior type H (**b**). The bottom row corresponds to all the independent runs of the EM learning algorithm with the prior type 1 (**c**) and the prior type H (**d**). The red colour highlights the curves with the maximum and minimum areas under curves

equates the MAP and MLE. We can conclude that the choice of the hyperparameter settings is not a problem for the EM learning algorithm and we can even simplify the derivations considering only the MLE without the prior influence.

The VB and GS learning algorithms require a proper choice of the hyperparameter settings as they can significantly change the anomaly detection performance. This choice can be performed empirically or with the type II maximum likelihood approach [15].

3.5.3.2 Predictive Likelihood Approximation Influence

In this section the influence of the type of the predictive likelihood approximation on the anomaly detection results is studied.

The average results for both datasets (Fig. 3.8) demonstrate that the type of the predictive likelihood approximation does not remarkably influence the anomaly detection performance. As the plug-in approximation requires less computational resources both in terms of time and memory (as there is no need to sample and store

Table 3.2 Mean area under precision-recall curves

Dataset	EM	VB	GS
QMUL	0.3166	0.3155	0.2970
Idiap	0.3759	0.3729	0.3643

posterior samples and average among them) this type of approximation is recommended to be used for anomaly detection in the proposed framework.

3.5.3.3 Learning Algorithms Comparison

This section compares the anomaly detection performance obtained by three learning algorithms.

The best results in terms of a mean area under a precision-recall curve are obtained by the EM learning algorithm, the worst results are obtained by the GS learning algorithm (Fig. 3.8 and Table 3.2). In Table 3.2 for each learning algorithm the group of hyperparameter settings and the type of predictive likelihood approximation is chosen to have the maximum of the mean area under curves, where a mean is taken over independent runs of the same method and maximum is taken among different settings for the same learning algorithm.

Figure 3.10 presents the best and the worst precision-recall curves (in terms of the area under them) for the individual runs of the learning algorithms. The figure shows that among the individual runs the EM learning algorithm also demonstrates the most accurate results. Although, the minimum area under the precision-recall curve for the EM learning algorithm is less than the area under the corresponding curve for the VB algorithm. This means that the variance among the individual curves for the EM learning algorithm is larger in comparison with the VB learning algorithm.

The variance of the precision-recall curves for both VB and GS learning algorithms are relatively small. However, the VB learning algorithm has the curves higher than the curves obtained by the GS learning algorithm. This can be confirmed by examination of the best and worst precision-recall curves (Fig. 3.10) and the mean values of the area under curves (Fig. 3.8 and Table 3.2).

3.5.3.4 Anomaly Localisation

We apply the proposed method for anomaly localisation, presented in Sect. 3.4.2, and get promising results. We demonstrate the localisation results for the EM learning algorithm with the prior type H+1 on both datasets in Fig. 3.11. It can be seen that the abnormal events correctly localised by the proposed method.

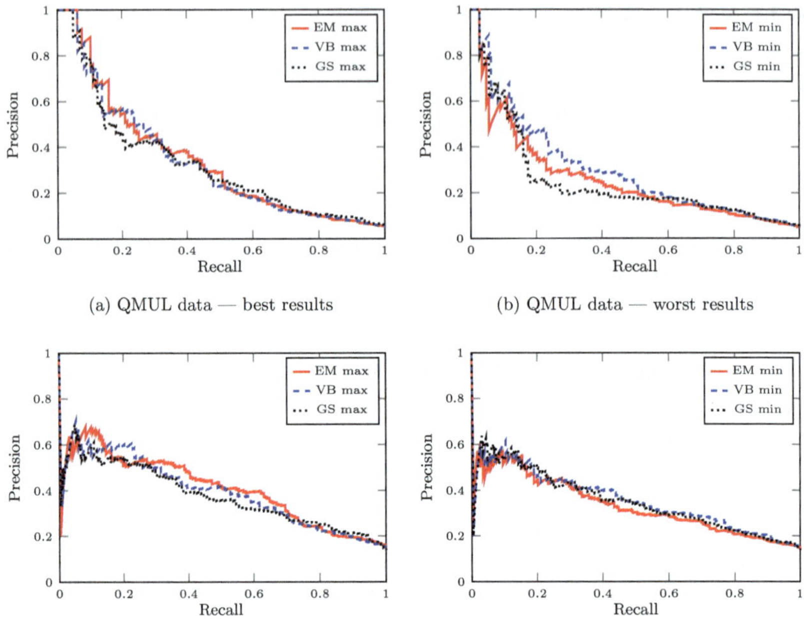

(a) QMUL data — best results

(b) QMUL data — worst results

(c) Idiap data — best results

(d) Idiap data — worst results

Fig. 3.10 Precision-recall curves with the maximum and minimum areas under curves for the three learning algorithms (maximum and minimum is among all the runs with different initialisations for all groups of hyperparameter settings and all types of predictive likelihood approximations). **a** presents the "best" curves for the QMUL data, i.e., the curves with the maximum area under a curve. **b** presents the "worst" curves for the QMUL data, i.e., the curves with the minimum area under a curve. **c** presents the "best" curves for the Idiap data, **d**—the "worst" curves for the Idiap data

3.6 Summary

This chapter presents two learning algorithms for the dynamic topic model for behaviour analysis in video: the EM-algorithm is developed for the MAP estimates of the model parameters and a variational Bayes inference algorithm is developed for calculating the posterior distributions of them. A detailed comparison of these proposed learning algorithms with the Gibbs sampling based algorithm developed in [2] is presented. The differences and the similarities of the theoretical aspects for all three learning algorithms are well emphasised. An empirical comparison is performed for abnormal behaviour detection using two unlabelled real video datasets. Both proposed learning algorithms demonstrate more accurate results than the algorithm proposed in [2] in terms of anomaly detection performance.

The EM learning algorithm demonstrates the best results in terms of the mean values of the performance measure, obtained by the independent runs of the algorithm with different random initialisations. Although, it is noticed that the variance among

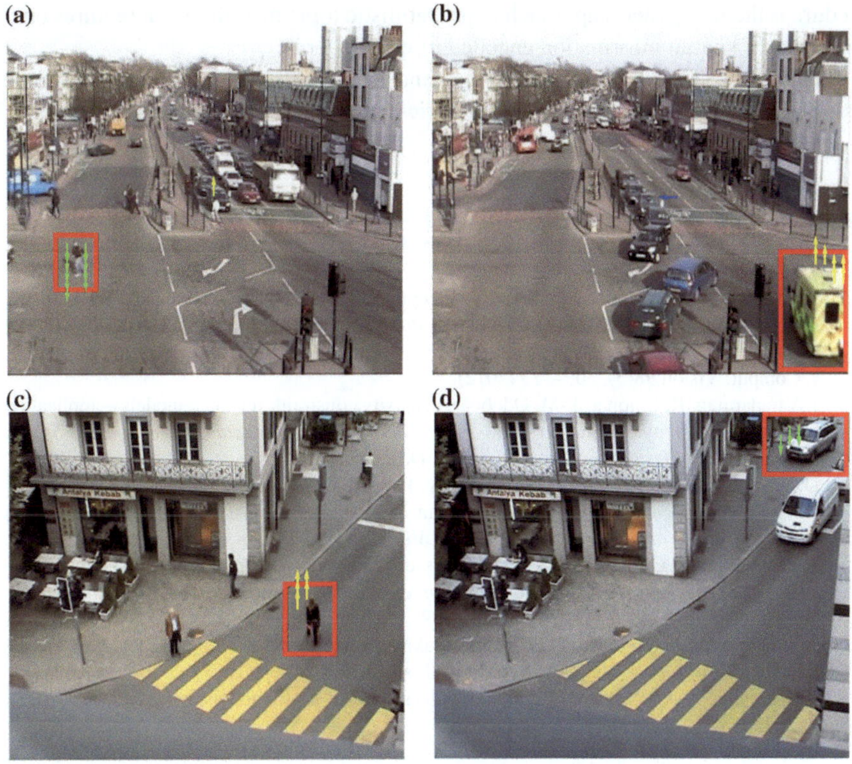

Fig. 3.11 Examples of anomalies localisation. The red rectangle is the manual localisation. The arrows represent the visual words with the smallest predictive likelihood, the locations of the arrows are the results of the algorithmic anomalies localisation

the precision-recall curves of the individual runs is relatively high. The variational Bayes learning algorithm shows the smaller variance among the precision-recall curves than the EM-algorithm and the VB algorithm answers are more robust to different initialisation values. However, the results of the algorithm are significantly influenced by the choice of the hyperparameters. The hyperparameters require additional tuning before the algorithm can be applied to data. Note that the results of the EM learning algorithm only slightly depend on the choice of the hyperparameter settings. Moreover, the hyperparameter can be even set to cancel prior influence on the estimates of parameters obtained by the EM algorithm that equates MLE and MAP estimates. Both proposed learning algorithms—EM and VB—provide more accurate results in comparison to the Gibbs sampling based algorithm.

We also demonstrate that consideration of predictive likelihoods of visual words rather than visual documents can provide satisfactory results about locations of anomalies within a frame. In our best knowledge the proposed localisation procedure is the first general approach in probabilistic topic modeling that requires only presence of spatial information encoded in visual words.

In the MCTM the number of topics and behaviours is limited and should be specified in advance. A novel nonparametric model is introduced in the next chapter.

References

1. X. Wang, X. Ma, Unsupervised activity perception in crowded and complicated scenes using hierarchical Bayesian models. IEEE Trans. Pattern Anal. Mach. Intell. **31**(3), 539–555 (2009)
2. T. Hospedales, S. Gong, T. Xiang, Video behaviour mining using a dynamic topic model. Int. J. Comput. Vision **98**(3), 303–323 (2012)
3. J. Varadarajan, R. Emonet, J.-M. Odobez, A sparsity constraint for topic models—application to temporal activity mining, in *NIPS-2010 Workshop on Practical Applications of Sparse Modeling: Open Issues and New Directions*, December 2010
4. D. Kuettel, M. Breitenstein, L. Van Gool, V. Ferrari, What's going on? Discovering spatio-temporal dependencies in dynamic scenes. in *Proceedings of the 2010 IEEE Conference on Computer Vision and Pattern Recognition (CVPR)*, pp. 1951–1958, June 2010
5. J. Varadarajan, J. Odobez, Topic models for scene analysis and abnormality detection, in *Proceedings of the 12th IEEE International Conference on Computer Vision Workshops (ICCV Workshops)*, pp. 1338–1345, September 2009
6. A.P. Dempster, N.M. Laird, D.B. Rubin, Maximum likelihood from incomplete data via the EM algorithm. J. Roy. Stat. Soc. Ser. B (Methodol.) **39**(1), 1–38 (1977)
7. K. Vorontsov, A. Potapenko, Additive regularization of topic models. Mach. Learn. **101**(1), 1–21 (2015)
8. C.M. Bishop, *Pattern Recognition and Machine Learning (Information Science and Statistics)* (Springer, New York, 2006)
9. M. Jordan, Z. Ghahramani, T. Jaakkola, L. Saul, An introduction to variational methods for graphical models. Mach. Learn. **37**(2), 183–233 (1999)
10. T.L. Griffiths, M. Steyvers, Finding scientific topics. Proc. Natl. Acad. Sci. **101**(suppl 1), 5228–5235 (2004)
11. J.S. Liu, W.H. Wong, A. Kong, Covariance structure of the Gibbs sampler with applications to the comparisons of estimators and augmentation schemes. Biometrika **81**(1), 27–40 (1994)
12. A. Asuncion, M. Welling, P. Smyth, Y.W. Teh, On smoothing and inference for topic models, in *Proceedings of the 25th Conference on Uncertainty in Artificial Intelligence (UAI 2009)*, pp. 27–34, June 2009
13. T.S.F. Haines, T. Xiang, Video topic modelling with behavioural segmentation, in *Proceedings of the 1st ACM International Workshop on Multimodal Pervasive Video Analysis*, pp. 53–58, October 2010
14. D. Pathak, A. Sharang, A. Mukerjee, Anomaly localization in topic-based analysis of surveillance videos, in *Proceedings of the 2015 IEEE Winter Conference on Applications of Computer Vision*, pp. 389–395, January 2015
15. K.P. Murphy, *Machine Learning: A Probabilistic Perspective* (MIT press, New York, 2012)

Chapter 4
Dynamic Hierarchical Dirichlet Process

In Chap. 3 the dynamic topic model and learning algorithms for it are considered. In that model the numbers of topics and topic mixtures are fixed. This chapter introduces a novel dynamic nonparametric topic model that allows a potentially infinite number of topics and, in practice, the number of topics is determined by the data. The application of the model for behaviour analysis and anomaly detection in video is considered in detail, however, the proposed model is not limited to this application (discussion of other potential application areas is presented in Sect. 6.2.4). In this chapter visual documents and words are defined in the same manner as in the previous chapter (Sect. 3.1).

There are two types of dynamics in the topic modeling literature (Sect. 2.3.4). In the proposed model the dynamics on topic mixtures in documents are considered. The model is designed to encourage neighbouring documents to have similar topic mixtures.

Imagine that there is an infinitely long video sequence. Motion patterns, which are typical for a scene, may appear and disappear and the total number of these patterns may be infinite. The motion patterns are modelled as topics in the topic model, hence the number of topics in the topic model may potentially be infinite. In real life this means that with the growth of a data size the number of topics is expected to increase. This intuition may be simulated by a nonparametric model [1].

In anomaly detection it is essential to make a decision as soon as possible to warn a human operator. Therefore, batch and online inference for the model based on the Gibbs sampler is developed in this chapter. During the batch offline set-up the Gibbs sampler processes a training set of documents, estimating distributions of words in topics. During the online set-up test documents are processed one by one. The main goal of the online inference is to estimate a topic mixture for the current document, without reconsideration of all the previous documents.

A final anomaly detection decision is based on a normality measure that is proposed here based on predictive likelihood of new data.

O. Isupova, *Machine Learning Methods for Behaviour Analysis and Anomaly Detection in Video*, Springer Theses, https://doi.org/10.1007/978-3-319-75508-3_4

The results of the work presented in this chapter are disseminated in:

- O. Isupova, D. Kuzin, L. Mihaylova. "Dynamic Hierarchical Dirichlet Process for Abnormal Behaviour Detection in Video", in *Proceedings of the 19th International Conference on Information Fusion*, 5–8 July 2016, Heidelberg, Germany, pp. 750–757
- O. Isupova, D. Kuzin, L. Mihaylova. "Anomaly Detection in Video with Bayesian Nonparametrics", in *ICML 2016 Anomaly detection Workshop*, 24 June 2016, New York, NY, USA

The remainder of the chapter is organised as follows. The hierarchical Dirichlet process topic model is overviewed in Sect. 4.1. The proposed model is described in Sect. 4.2. Section 4.3 presents the inference for the model, while Sect. 4.4 introduces the anomaly detection procedure. The experimental results are given in Sect. 4.5. Section 4.6 summarises the chapter.

4.1 Hierarchical Dirichlet Process Topic Model

In contrast to the model from Chap. 3 and basic parametric topic models, such as LDA and PLSA (Sect. 2.3.2), we consider a nonparametric set-up where the number of topics is not limited: $\{\phi_k\}_{k=1:\infty}$. Moreover, it is assumed that observing an infinite amount of data we can expect to have an infinite number of topics.

This kind of mixture models with a potentially infinite number of mixture components can be modelled with the hierarchical Dirichlet process (HDP) [2]. The HDP is a hierarchical extension of the Dirichlet process (DP), which is a distribution over random distributions [3]. Each document $\mathbf{w}_j$ is associated with a sample G_j from a DP:

$$G_j \sim \mathcal{DP}(\alpha, G_0), \tag{4.1}$$

where $\mathcal{DP}(\cdot, \cdot)$ denotes a DP, α is a concentration parameter, G_0 is a base measure. The sample G_j can be seen as a vector of mixture component weights, where the number of components is infinite.

The base measure G_0 itself is a sample from another DP:

$$G_0 \sim \mathcal{DP}(\gamma, H), \tag{4.2}$$

with the concentration parameter γ and the base measure H. This shared measure G_0 from a DP ensures that the documents will have the same set of topics but with different weights. Indeed, G_0 is almost surely discrete [4], concentrating its mass on the atoms $\boldsymbol{\phi}_k$ drawn from H. Therefore, G_j picks the mixture components from this set of atoms.

A topic, that is an atom $\boldsymbol{\phi}_k$, is modelled as the categorical distribution with a probability ϕ_{wk} of choosing a word w. The base measure H is therefore chosen as the conjugate Dirichlet distribution, where $\boldsymbol{\eta}$ denotes a parameter of this Dirichlet distribution. As usual a symmetric Dirichlet parameter is considered: throughout this chapter $\boldsymbol{\eta} = \{\eta, \ldots, \eta\}$ is used for simplicity.

Document j is formed by repeating the procedure of drawing a topic from the mixture:

$$\theta_{ji} \sim G_j \tag{4.3}$$

and drawing a word from the chosen topic for every token i:

$$w_{ji} \sim Cat(\theta_{ji}) \tag{4.4}$$

4.1.1 Chinese Restaurant Franchise

There are several versions of the HDP representation (as well as the DP) [2]. In this work the representation called Chinese restaurant franchise is considered, as it is used for the derivation of the Gibbs sampling inference scheme. In this metaphor, each document corresponds to a "restaurant"; words correspond to "customers" of the restaurant. The words in the documents are grouped around "tables" and the number of tables in each document is unlimited. Each table serves a "dish", which corresponds to a topic. The "menu" of dishes, i.e., the set of the topics, is shared among all the restaurants.

Let t_{ji} denote a table assignment for token i in document j and let k_{jt} denote a topic assignment for table t in document j. Let n_{jt} denote the number of words assigned to table t in document j and m_{jk} denote the number of tables in document j serving topic k. The dots in subscripts mean marginalisation over the corresponding dimension, e.g., $m_{\cdot k}$ denotes the number of tables among all the documents serving topic k, while $m_{j\cdot}$ denotes the total number of tables in document j. Marginalisation over both dimensions $m_{\cdot\cdot}$ means the total number of tables in the dataset.

The data generative process is as follows. A new token comes to document j and chooses one of the occupied tables with a probability proportional to the number of words n_{jt} assigned to this table, or the new token starts a new table with a probability proportional to α (Fig. 4.1a):

$$p(t_{ji} = t | t_{j1}, \ldots, t_{ji-1}, \alpha) = \begin{cases} \dfrac{n_{jt}}{i - 1 + \alpha}, & \text{if } t = 1 : m_{j\cdot}; \\ \dfrac{\alpha}{i - 1 + \alpha}, & \text{if } t = t^{\text{new}}. \end{cases} \tag{4.5}$$

If the token starts a new table, it chooses a topic for it. This topic will be shared by all tokens that may join this table later. The process of choosing a topic for a table can also be formulated as customer-to-table assignment in a top-level restaurant, where customers are tables from document-level restaurants (Fig. 4.1b). This top-level restaurant shows global topic popularity.

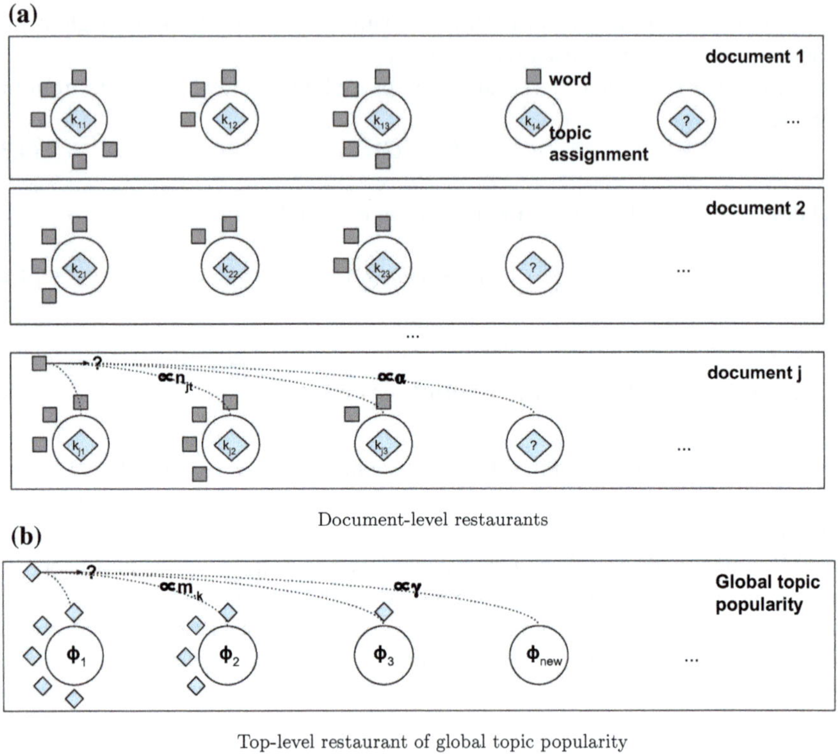

Fig. 4.1 Chinese restaurant franchise for the HDP

Therefore, the token that starts a new table in a document-level restaurant chooses one of the used topics with a probability proportional to the number of tables $m_{.k}$ that serve this topic among all the documents, or the token chooses a new topic, sampling it from the base measure H, with a probability proportional to γ:

$$p(k_{jt^{\text{new}}} = k | k_{11}, \ldots, k_{jt-1}, \gamma) = \begin{cases} \dfrac{m_{.k}}{m_{..} + \gamma}, & \text{if } k = 1 : K; \\ \dfrac{\gamma}{m_{..} + \gamma}, & \text{if } k = k^{\text{new}}, \end{cases} \tag{4.6}$$

where K is the number of topics used so far.

Once the token is assigned to the table t_{ji} with the topic $k_{jt_{ji}}$, the word w_{ji} for this token is sampled from this topic:

$$w_{jt} \sim Cat(\phi_{k_{jt_{ji}}}). \tag{4.7}$$

The correspondence between two representations of the HDP (4.1)–(4.4) and (4.5)–(4.7) is based on the following equality: $\theta_{ji} = \phi_{k_{jt_{ji}}}$.

4.2 Proposed Dynamic Hierarchical Dirichlet Process Topic Model

In the HDP, exchangeability of documents and words is assumed, which means that the joint probability of the data is independent of the order of documents and words in documents. However, in video processing applications this assumption may be invalid. While the words inside the documents are still exchangeable, the documents themselves are not. All actions and motions in real life last for some time, and it is expected that the topic mixture in the current document is similar to the topic mixture in the previous document. Some topics may appear and disappear but the core structure of the mixture component weights only slightly changes from document to document.

We propose a dynamic extension of the HDP topic model to take into account this intuition. In this model the probability of topic k explicitly depends on the usage of this topic in the current and previous documents $m_{jk} + m_{j-1k}$, therefore the topic distribution in the current document would be similar to the topic distribution in the previous document. The topic probability still additionally depends on the number of tables that serve this topic in the whole dataset $m_{\cdot k}$, but this number is weighted by a non-negative value λ, which is a parameter of the model. As in the previous case, it is possible to sample a new topic from the base measure H.

The generative process can be then formulated as follows. A new token comes to a document and, as before, chooses one of the occupied tables t with a probability proportional to the number of words n_{jt} already assigned to it, or it starts a new table with a probability proportional to the parameter α (Fig. 4.2a):

$$p(t_{ji} = t | t_{j1}, \ldots, t_{ji-1}, \alpha) = \begin{cases} \dfrac{n_{jt}}{i - 1 + \alpha}, & \text{if } t = 1 : m_{j\cdot}; \\ \dfrac{\alpha}{i - 1 + \alpha}, & \text{if } t = t^{\text{new}}. \end{cases} \tag{4.8}$$

The token that starts a new table should choose a topic for this table. One of the used topics k is chosen with a probability proportional to the sum of the number of tables having this topic in the current and previous documents $m_{jk} + m_{j-1k}$ and the weighted number of tables among all the previously observed documents, which have this topic, $\lambda m_{\cdot k}$. A new topic can be chosen for table t with a probability proportional to the parameter γ (Fig. 4.2b):

$$p(k_{jt} = k | k_{11}, \ldots, k_{jt-1}, \gamma) = \begin{cases} \dfrac{m_{jk} + m_{j-1k} + \lambda m_{\cdot k}}{m_{j\cdot} + m_{j-1\cdot} + \lambda m_{\cdot\cdot} + \gamma}, & \text{if } k = 1 : K; \\ \dfrac{\gamma}{m_{j\cdot} + m_{j-1\cdot} + \lambda m_{\cdot\cdot} + \gamma}, & \text{it } k = k^{\text{new}}. \end{cases} \tag{4.9}$$

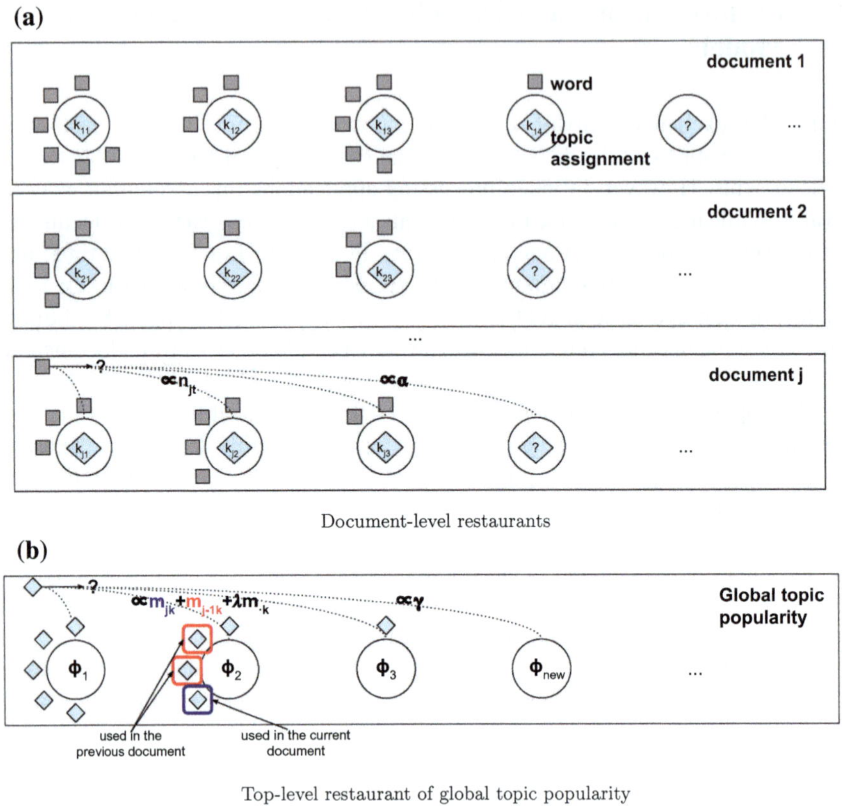

Fig. 4.2 Chinese restaurant franchise for the dynamic HDP

Finally, word w_{ji} is sampled for token i in document j, assigned to the table $t_{ji} = t$, which serves the topic $k_{jt} = k$. The word is sampled from the corresponding topic k:

$$w_{ji} \sim Cat(\phi_k). \tag{4.10}$$

4.3 Inference

Standard inference algorithms process an entire dataset. For large or stream datasets this batch set-up is computationally intractable. Online algorithms process data in a sequential manner, one data point at a time, incrementally updating the variables, corresponding to the whole dataset. It allows to save memory space and reduce the computational time. In this chapter combination of offline and online inference is proposed and this section describes it in detail.

The Gibbs sampling scheme is used. The inference procedure consists of two parts. Firstly, the traditional batch set-up of the Gibbs sampling is applied to the training set of documents $\mathbf{w}_{1:J_{tr}}$. Then an online set-up for the inference is applied for the test documents $\mathbf{w}_j, j > J_{tr}$. This means that the information about a test document is incrementally added to the model, not requiring to process the training documents again.

In the Gibbs sampling inference scheme the hidden variables t_{ji} and k_{jt} are sampled from their conditional distributions. In the Gibbs sampler for the HDP model, exchangeability of documents and words is used by treating the current variable t_{ji} as the table assignment for the last token in the last document and k_{jt} as the topic assignment for the last table in the last document. There is no exchangeability of documents in the proposed model, but words inside a document are still exchangeable. Therefore, the variable t_{ji} can be treated as the table assignment for the last token in the current document j, and the variable k_{jt} can be treated as the topic assignment for the last table in the current document j. The documents are processed in the order they appear in the dataset.

The following notation is used below. Let V denote the size of the word vocabulary ($V = |\mathcal{V}|$), $\mathbf{t}_{j_1:j_2} = \{t_{ji}\}_{j=j_1:j_2, i=1:N_j}$ is the set of the table assignments for all the tokens in the documents from j_1 to j_2. Let $\mathbf{k}_{j_1:j_2} = \{k_{jt}\}_{j=j_1:j_2, t=1:m_{j\cdot}}$ denote the corresponding set for the topic assignments. Let $m_{j_1:j_2\,k}$ denote the number of tables having topic k in the documents from j_1 to j_2. Let also $\mathbf{w}_{jt} = \{w_{ji}\}_{i=1:N_j, t_{ji}=t}$ denote the words assigned to table t in the document j.

Recall that l_{wk} denotes the number of times word w is associated with topic k, let $l_{\cdot k}$ denote the number of tokens associated with topic k: $l_{\cdot k} = \sum_w l_{wk}$, regardless of the word assignments. The notation $l_{wk}^{j_1:j_2}$ is used for the number of times word w is associated with topic k in the documents from j_1 to j_2.

The superscript $-ji$ indicates the corresponding variable without considering token i in document j, e.g., the set variable $\mathbf{t}_{j_1:j_2}^{-ji} = \mathbf{t}_{j_1:j_2} \setminus \{t_{ji}\}$ or the count n_{jt}^{-ji} is the number of words assigned to table t in document j excluding the word for token i. Similarly, the superscript $-jt$ means the corresponding variable without considering table t in document j.

4.3.1 Batch Collapsed Gibbs Sampling

4.3.1.1 Sampling Topic Assignment k_{jt}

The topic assignment k_{jt} for table t in document j is sampled from the conditional distribution given the observed data $\mathbf{w}_{1:J_{tr}}$ and all the other hidden variables, i.e., the table assignments for all the tokens $\mathbf{t}_{1:J_{tr}}$ and the topic assignments for all the other tables $\mathbf{k}_{1:J_{tr}}^{-jt}$:

$$p(k_{jt} = k|\mathbf{w}_{1:J_{tr}}, \mathbf{t}_{1:J_{tr}}, \mathbf{k}^{-jt}_{1:J_{tr}}) \propto p(\mathbf{w}_{jt}|k_{jt} = k, \mathbf{k}^{-jt}_{1:J_{tr}}, \mathbf{t}_{1:J_{tr}}, \mathbf{w}^{-jt}_{1:J_{tr}})\, p(k_{jt} = k|\mathbf{k}^{-jt}_{1:J_{tr}}). \tag{4.11}$$

The likelihood term $p(\mathbf{w}_{jt}|k_{jt} = k, \mathbf{k}^{-jt}_{1:J_{tr}}, \mathbf{t}_{1:J_{tr}}, \mathbf{w}^{-jt}_{1:J_{tr}})$ can be computed by integrating out the distribution ϕ_k:

$$\mathsf{f}^{-jt}(\mathbf{w}_{jt}) \stackrel{\text{def}}{=} p(\mathbf{w}_{jt}|k_{jt} = k, \mathbf{k}^{-jt}_{1:J_{tr}}, \mathbf{t}_{1:J_{tr}}, \mathbf{w}^{-jt}_{1:J_{tr}}) =$$
$$\int p(\mathbf{w}_{jt}|\phi_k)\, p(\phi_k|\mathbf{k}^{-jt}_{1:J_{tr}}, \mathbf{t}_{1:J_{tr}}, \mathbf{w}^{-jt}_{1:J_{tr}}) \mathrm{d}\phi_k = \frac{\prod_{w\in\mathcal{V}} \Gamma(l_{wk} + \eta)}{\Gamma(l_{\cdot k} + V\eta)} \frac{\Gamma(l^{-jt}_{\cdot k} + V\eta)}{\prod_{w\in\mathcal{V}} \Gamma(l^{-jt}_{wk} + \eta)}, \tag{4.12}$$

where $\Gamma(\cdot)$ is the gamma function. In the case when k is a new topic ($k = k^{\text{new}} = K + 1$) the integration is performed over the prior distribution for $\phi_{k^{\text{new}}}$. The obtained likelihood term (4.12) is then:

$$\mathsf{f}^{-jt}_{k^{\text{new}}}(\mathbf{w}_{jt}) = \frac{\prod_{w\in\mathcal{V}} \Gamma(l_{wk^{\text{new}}} + \eta)}{\Gamma(l_{\cdot k^{\text{new}}} + V\eta)} \frac{\Gamma(V\eta)}{(\Gamma(\eta))^V}. \tag{4.13}$$

The second multiplier in (4.11) $p(k_{jt} = k|\mathbf{k}^{-jt})$ can be further factorised as:

$$p(k_{jt} = k|\mathbf{k}^{-jt}_{1:J_{tr}}) \propto p(\mathbf{k}_{j+1:J}|\mathbf{k}^{-jt}_{1:j}, k_{jt} = k)\, p(k_{jt} = k|\mathbf{k}^{-jt}_{1:j}). \tag{4.14}$$

The first term in (4.14) is the probability of the topic assignments for all the tables in the next documents depending on the change of the topic assignment for table t in document j. Consider the topic assignments in document $j + 1$ first. From (4.9) it is:

$$g^{-jt}_k(\mathbf{k}_{j+1}) \stackrel{\text{def}}{=} p(\mathbf{k}_{j+1}|\mathbf{k}^{-jt}_{1:j}, k_{jt} = k) =$$
$$\frac{\gamma^{|\mathcal{K}^{\text{born}}_{j+1}|} \prod_{\acute{k}\in\mathcal{K}^{\text{born}}_{j+1}} \left(m_{j+1\acute{k}} - 1\right)! \,(1 + \lambda)^{m_{j+1l}-1}}{\prod_{\acute{t}=1}^{m_{j+1\cdot}} \left(m_{j\cdot} + \acute{t} - 1 + \lambda\left(m_{1:j\cdot} + \acute{t} - 1\right) + \gamma\right)} \times$$
$$\prod_{\acute{k}\notin\mathcal{K}^{\text{born}}_{j+1}} \prod_{\acute{t}=1}^{m_{j+1\acute{k}}} \left(m^{-jt\to k}_{j\acute{k}} + \acute{t} - 1 + \lambda\left(m^{-jt\to k}_{1:j\acute{k}} + \acute{t} - 1\right)\right) \propto$$
$$\prod_{\acute{k}\notin\mathcal{K}^{\text{born}}_{j+1}} \prod_{\acute{t}=1}^{m_{j+1\acute{k}}} \left(m^{-jt\to k}_{j\acute{k}} + \acute{t} - 1 + \lambda\left(m^{-jt\to k}_{1:j\acute{k}} + \acute{t} - 1\right)\right), \tag{4.15}$$

where the sign of proportionality is used with respect to k_{jt}, $\mathcal{K}_{j+1}^{\text{born}}$ is the set of the topics that first appear in document $j+1$, the superscript $-jt \to k$ means that k_{jt} is set to k for the corresponding counts, $|\cdot|$ is the cardinality of the set.

The similar probabilities of the topic assignments for all the next documents $j' = j+2 : J_{tr}$ depend on k only in the term $m_{1:j'-1\cdot}^{-jt\to k}$. It is assumed that the influence of k on these probabilities is not significant and the first term in (4.14) is approximated by the probability of the topic assignments (4.15) in document $j+1$ only:

$$p(\mathbf{k}_{j+1:J}|\mathbf{k}_{1:j}^{-jt}, k_{jt} = k) \approx g_k^{-jt}(\mathbf{k}_{j+1}). \tag{4.16}$$

The second term in (4.14) is the prior for k_{jt}:

$$p(k_{jt} = k|\mathbf{k}_{1:j}^{-jt}) \propto \begin{cases} m_{jk}^{-jt} + m_{j-1k} + \lambda m_{1:j\,k}^{-jt}, & \text{if } k = 1 : K; \\ \gamma, \text{ if } k = k^{\text{new}}. \end{cases} \tag{4.17}$$

Substituting (4.16) and (4.17) into (4.14), $p(k_{jt} = k|\mathbf{k}_{1:J_{tr}}^{-jt})$ is computed as follows:

$$p(k_{jt} = k|\mathbf{k}_{1:J_{tr}}^{-jt}) \propto \begin{cases} g_k^{-jt}(\mathbf{k}_{j+1})(m_{jt} + m_{j-1k} + \lambda m_{1:j\,k}), & \text{if } k = 1 : K; \\ g_{k^{\text{new}}}^{-jt}(\mathbf{k}_{j+1})\gamma, \text{ if } k = k^{\text{new}}. \end{cases} \tag{4.18}$$

The topic assignment sampling distribution can be then expressed as:

$$p(k_{jt} = k|\mathbf{w}_{1:J_{tr}}, \mathbf{t}_{1:J_{tr}}, \mathbf{k}_{1:J_{tr}}^{-jt}) \propto \mathsf{f}_k^{-jt}(\mathbf{w}_{jt})\, p(k_{jt} = k|\mathbf{k}_{1:J_{tr}}^{-jt}), \tag{4.19}$$

where $\mathsf{f}_k^{-jt}(\mathbf{w}_{jt})$ is given by (4.12)–(4.13) and $p(k_{jt} = k|\mathbf{k}_{1:J_{tr}}^{-jt})$ is given by (4.18).

4.3.1.2 Sampling t_{ji}

The table assignment t_{ji} for token i in document j is sampled from the conditional distribution given the observed data $\mathbf{w}_{1:J_{tr}}$ and all the other hidden variables, i.e., the topic assignments for all the tables $\mathbf{k}_{1:J_{tr}}$ and the table assignments for all the other tokens $\mathbf{t}_{1:J_{tr}}^{-ji}$:

$$p(t_{ji} = t|\mathbf{w}_{1:J_{tr}}, \mathbf{k}_{1:J_{tr}}, \mathbf{t}_{1:J_{tr}}^{-ji}) \propto p(w_{ji}|\mathbf{t}_{1:J_{tr}}^{-ji}, t_{ji} = t, \mathbf{w}_{1:J_{tr}}^{-ji}, \mathbf{k}_{1:J_{tr}})\, p(t_{ji} = t|\mathbf{t}_{1:J_{tr}}^{-ji}) \tag{4.20}$$

The first term in (4.20) is the likelihood of word w_{ji}. It depends on whether t is one of the previously used table or it is a new table. For the case when t is the table, which is already used, the likelihood is:

$$\mathsf{f}_{k_{jt}}^{-ji}(w_{ji}) = p(w_{ji}|t_{ji} = t, \mathbf{t}_{1:J_{tr}}^{-ji}, \mathbf{k}_{1:J_{tr}}, \mathbf{w}_{1:J_{tr}}^{-ji}) = \frac{l_{w_{ji}\,k_{jt}} + \eta}{l_{\cdot\,k_{jt}} + V\eta} \tag{4.21}$$

Consider now the case when $t_{ji} = t^{\text{new}}$, i.e., the likelihood of the word w_{ji} being assigned to a new table. This likelihood can be found by integrating out the possible topic assignments $k_{jt^{\text{new}}}$ for this table:

$$r_{t^{\text{new}}}(w_{ji}) \stackrel{\text{def}}{=} p(w_{ji}|\mathbf{t}_{1:J_{tr}}^{-ji}, t_{ji} = t^{\text{new}}, \mathbf{w}_{1:J_{tr}}^{-ji}, \mathbf{k}_{1:J_{tr}}) = \sum_{k=1}^{K} \mathsf{f}_k^{-ji}(w_{ji})\, p(k_{jt^{\text{new}}} = k|\mathbf{k}_{1:J_{tr}}) + \mathsf{f}_{k^{\text{new}}}^{-ji}(w_{ji})\, p(k_{jt^{\text{new}}} = k^{\text{new}}|\mathbf{k}_{1:J_{tr}}), \tag{4.22}$$

where $p(k_{jt^{\text{new}}} = k|\mathbf{k}_{1:J_{tr}})$ is given by (4.18).

The second term in (4.20) is the prior for t_{ji}:

$$p(t_{ji} = t|\mathbf{t}_{1:J_{tr}}^{-ji}) \propto \begin{cases} n_{jt}, & \text{if } t = 1 : m_{j\cdot}; \\ \alpha, & \text{if } t = t^{\text{new}}. \end{cases} \tag{4.23}$$

Then the conditional distribution for sampling a table assignment t_{ji} is:

$$p(t_{ji} = t|\mathbf{w}_{1:J_{tr}}, \mathbf{k}_{1:J_{tr}}, \mathbf{t}_{1:J_{tr}}^{-ji}) \propto \begin{cases} \mathsf{f}_{k_{jt}}^{-ji}(w_{ji}) n_{jt}, & \text{if } t = 1 : m_{j\cdot}; \\ r_{t^{\text{new}}}(w_{ji})\alpha, & \text{if } t = t^{\text{new}}. \end{cases} \tag{4.24}$$

If a new table is sampled, then a topic for it is drawn from (4.19).

4.3.2 *Online Inference*

In online or distributed implementations of inference algorithms in topic modeling global variables (that depend on the whole set of data) are separated from local variables (that depend only on the current document) [5–7].

For the proposed dynamic HDP model the global variables are the distributions ϕ_k, which are approximated by the counts l_{wk}, and the global topic popularity, which is estimated by the counts $m_{\cdot k}$. Note that the relative relationship between counts is important, rather than the absolute values of the counts. The local variables are the topic mixture weights for each document, governed by the counts m_{jk}. The training dataset is assumed to be large enough such that the global variables are well estimated by the counts available during the training stage and a new document can only slightly change the obtained ratios of the counts.

Following this assumption, the learning procedure is organised as follows. The batch Gibbs sampler is run for the training set $\mathbf{w}_{1:J_{tr}}$ of the documents. After this training stage the global counts l_{wk} and $m_{\cdot k}$ for all w and k are stored and used for the online inference of the test documents: $\mathbf{w}_j, j > J_{tr}$. For each test document $\mathbf{w}_j$ the online Gibbs sampler is run to sample table assignments and topic assignments for this document only. The online Gibbs sampler updates the local counts m_{jk}. After the Gibbs sampler converges, the global counts l_{wk} and $m_{\cdot k}$ are updated with the information obtained by the new document.

The equations for the online version of the Gibbs sampler differ from the batch ones in the update formula for k_{jt}. Namely, the conditional probability $p(k_{jt} = k|\mathbf{k}_{1:j}^{-jt})$ in the topic assignment sampling distribution (4.19) differs from (4.14). As successive documents are not observed while processing the current document, this probability consists only of the prior term $p(k_{jt} = k|\mathbf{k}_{1:j}^{-jt})$:

$$p_{\text{online}}(k_{jt} = k|\mathbf{k}_{1:j}^{-jt}) = \begin{cases} m_{jk}^{-jt} + m_{j-1k} + \lambda m_{1:j\,k}^{-jt}, & \text{if } k = 1:K; \\ \gamma, & \text{if } k = k^{\text{new}}. \end{cases} \tag{4.25}$$

Substituting this expression into (4.19) the obtained sampling distribution for the topic assignment in the online Gibbs sampler is:

$$p_{\text{online}}(k_{jt} = k|\mathbf{w}_{1:j}, \mathbf{t}_{1:j}, \mathbf{k}_{1:j}^{-jt}) \propto \begin{cases} \mathrm{f}_k^{-jt}(\mathbf{w}_{jt})(m_{jt} + m_{j-1k} + \lambda m_{1:j\,k}), & \text{if } k = 1:K; \\ \mathrm{f}_{k^{\text{new}}}^{-jt}(\mathbf{w}_{jt})\,\gamma, & \text{if } k = k^{\text{new}}. \end{cases} \tag{4.26}$$

The updating distribution for the topic assignment in the online Gibbs sampler remains the same as in the batch version (4.24).

The similar idea of online inference is used in the sequential Markov chain Monte Carlo scheme [8]. In that approach a set of posterior samples obtained from the previous observations are used instead of one that is used in the proposed algorithm. Using several posterior samples can increase accuracy but it also multiplies computational complexity of processing each document by the number of posterior samples.

4.4 Anomaly Detection

Topic models provide a probabilistic framework for anomaly detection. Under this framework a normality measure is the likelihood of data.

The Gibbs sampler provides estimates of the distributions ϕ_k and posterior samples of the table and topic assignments. This information can be used to estimate the predictive likelihood of a new clip. The predictive likelihood, normalised by the length N_j of the clip in terms of visual words, is used as a normality measure in this chapter.

The predictive likelihood is estimated via a harmonic mean [9], as it allows to use the information from the posterior samples:

$$p(\mathbf{w}_j|\mathbf{w}_{1:j-1}) = \left(\sum_{\mathbf{t}_{1:j},\mathbf{k}_{1:j}} \frac{p(\mathbf{t}_{1:j}, \mathbf{k}_{1:j}|\mathbf{w}_j, \mathbf{w}_{1:j-1})}{p(\mathbf{w}_j|\mathbf{t}_{1:j}, \mathbf{k}_{1:j}, \mathbf{w}_{1:j-1})} \right)^{-1} \approx \left(\frac{1}{S} \sum_{s=1}^{S} \frac{1}{p(\mathbf{w}_j|\mathbf{t}^s_{1:j}, \mathbf{k}^s, \mathbf{w}_{1:j-1})} \right)^{-1}, \tag{4.27}$$

where S is the number of the posterior samples, $\mathbf{t}^s_{1:j}$ and $\mathbf{k}^s_{1:j}$ are from the s-th posterior sample obtained by the Gibbs sampler, and

$$p(\mathbf{w}_j|\mathbf{t}^s_{1:j}, \mathbf{k}^s, \mathbf{w}_{1:j-1}) = \prod_{k=1}^{K} \frac{\prod_{w\in\mathcal{V}} \Gamma(l^{1:j\,s}_{wk} + \eta)}{\Gamma(l^{1:j\,s}_{\cdot k} + V\eta)} \frac{\Gamma(l^{1:j-1\,s}_{\cdot k} + V\eta)}{\prod_{w\in\mathcal{V}} \Gamma(l^{1:j-1\,s}_{wk} + \eta)}. \tag{4.28}$$

The superscript s on the counts means these counts are from the s-th posterior sample.

The anomaly detection procedure is then as follows. The batch Gibbs sampler is run on the training dataset. Then for each clip from the test dataset first the online Gibbs sampler is run to obtain the posterior samples of the hidden variables corresponding to the current clip. Afterwards the normality measure:

$$\mathcal{A}(\mathbf{w}_j) = \frac{1}{N_j} \log p(\mathbf{w}_j|\mathbf{w}_{1:j-1}) \tag{4.29}$$

is computed for the current clip. If the normality measure is below a threshold, the clip is labelled as abnormal, otherwise as normal. And the next clip from the test dataset is processed.

4.5 Experiments

In this section the proposed method is applied to behaviour analysis and anomaly detection. The method is compared with the one, based on the HDP topic model, where for the HDP topic model the online version of the Gibbs sampler and the normality measure are derived similarly to the dynamic HDP. The methods are compared on both synthetic and real data. The area under precision-recall curves, introduced in Sect. 3.5.1, is used as a performance measure.

4.5.1 Synthetic Data

The popular "bar" data [9] is used as a synthetic dataset. In this data the vocabulary consists of $V = 25$ words, organised into a 5×5 matrix. There are 10 topics in total, the word distributions ϕ_k of these topics form vertical and horizontal bars in the matrix (Fig. 4.3).

The training dataset consisting of 2000 documents is generated from the proposed model (4.8)–(4.10), where 1% noise is added to the distributions ϕ_k. Each of the documents has 20 words. The hyperparameters are set to the following values for the generation: $\alpha = 1.5$, $\gamma = 2$, $\lambda = 0.5$.

Similarly, a test dataset consisting of 1000 documents is used, but where 300 random documents are generated as "abnormal". In the proposed model it is assumed that topic mixtures in neighbouring documents are similar. In contrast to this assumption, topics for an abnormal document are chosen uniformly from the set of all the topics except those used in the previous document.

Both algorithms are run for these datasets, computing the normality measure for all the test documents. The hyperparameters α, γ and λ are set to the same values as for the generation, $\eta = 0.2$ (η is not used in generation as the word distributions in topics are set manually). Each of the algorithms has 5 runs with different initialisations to obtain 5 independent posterior samples. Both batch and online samplers are run for 1000 burn-in iterations.

In Fig. 4.4 the precision-recall curves for the obtained normality measures are presented together with the precision-recall curve for the "true" normality measure. The "true" normality measure is computed using the likelihood based on the true

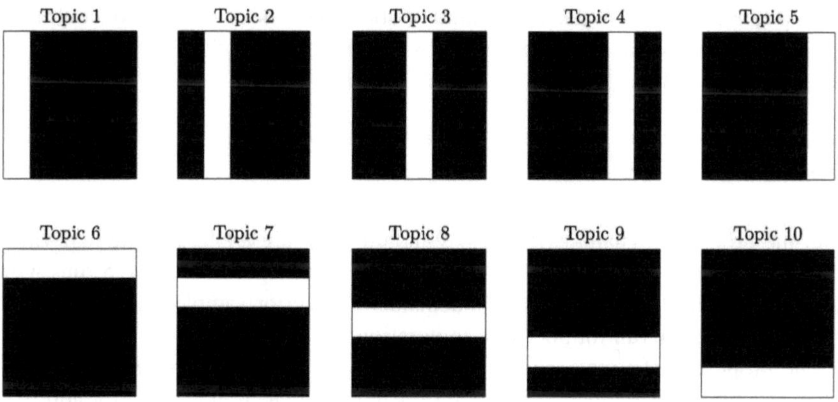

Fig. 4.3 Graphical representation of the topics in the synthetic dataset. There are 25 words, organised into a 5×5 matrix, where a word corresponds to a cell in this greyscale matrix. The topics are represented as the greyscale matrices, where the intensity of the cell indicates the probability of the corresponding word in a given topic, the lighter the colour the higher the probability value. For the presented topics black cells indicate zero probabilities of corresponding words, white cells represent probabilities equal to 0.2 of corresponding words

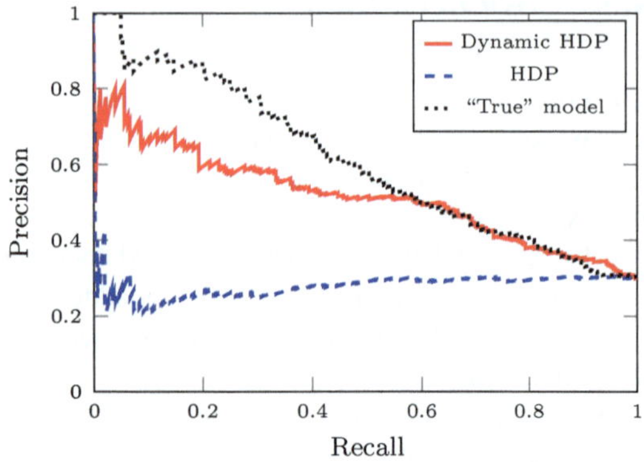

Fig. 4.4 Precision-recall curves for the synthetic data obtained by both models. The precision-recall curve, obtained by the likelihood, computed with the known true hidden variables, is labelled as a "true" model

Table 4.1 Dynamic HDP versus standard HDP. Area under precision-recall curves results

Dataset	Dynamic HDP	HDP	"True" model
Synthetic	0.5147	0.2817	0.6046
QMUL	0.3232	0.0980	–
Idiap	0.3542	0.2586	–

distributions ϕ_k and the true table and topic assignments $\mathbf{t}_{1:j}$ and $\mathbf{k}_{1:j}$, i.e., it corresponds to the model that can perfectly restore the latent variables. Table 4.1 contains the obtained values for the area under precision-recall curves.

The results show that the proposed dynamic HDP can detect the simulated abnormalities and its performance is competitive to the "true" model. The original HDP method is not expected to detect this kind of abnormalities, as they do not contradict its generative model, and it is confirmed by the empirical results.

4.5.2 *Real Video Data*

Both the proposed dynamic HDP and standard HDP are applied to the real video data—QMUL and Idiap, introduced in Sect. 3.5. Both algorithms have 5 runs with different initialisations to obtain 5 independent posterior samples. The batch and online samplers are run for 500 burn-in iterations.

The proposed inference scheme based on online processing of test documents achieves a fast decision making procedure. Anomaly detection is made for approximately 0.5427 s per visual document by the proposed dynamic HDP method with 5 posterior samples obtained by independent Gibbs samplers using 500 burn-in iterations.[1]

[1]The computational time is provided for a laptop computer with Intel Core i5 CPU with 2.4 GHz, 8 GB RAM using C++ implementation.

The hyperparameters for the dynamic HDP model are set as follows: $\alpha = 0.001$, $\gamma = 0.001$ and $\eta = 0.5$ on both datasets and $\lambda = 0.0005$ for the QMUL data and $\lambda = 0.5$ for the Idiap data. The standard HDP algorithm is used with the following settings: $\alpha = 0.001$, $\gamma = 0.001$ and $\eta = 0.5$ for both datasets.

Figure 4.5 shows visualisation of the topics learnt by the dynamic HDP for the QMUL data. The topics represent clear semantic motion patterns. Figure 4.5c and d demonstrate that topics can have overlapping subsets of visual words that in combination with other words form different topics. In this example the right turn can be a part of vertical traffic flow or can be observed in the scene together with the left turn.

Sample topics extracted from the Idiap data by the dynamic HDP algorithm are given in Fig. 4.6. The examples show that topics in the dynamic HDP can represent both global motion patterns such as traffic flows and local activities such as pedestrian motions in different parts of the scene.

The precision-recall curves for anomaly detection obtained by both algorithms are presented in Fig. 4.7. The corresponding values for the area under the curves can be found in Table 4.1. The results provided in Fig. 4.7 and Table 4.1 show that consideration of the dynamics proposed in this chapter for the HDP topic model significantly improves the performance of the algorithm in terms of anomaly detection. The improvement is achieved on both real datasets.

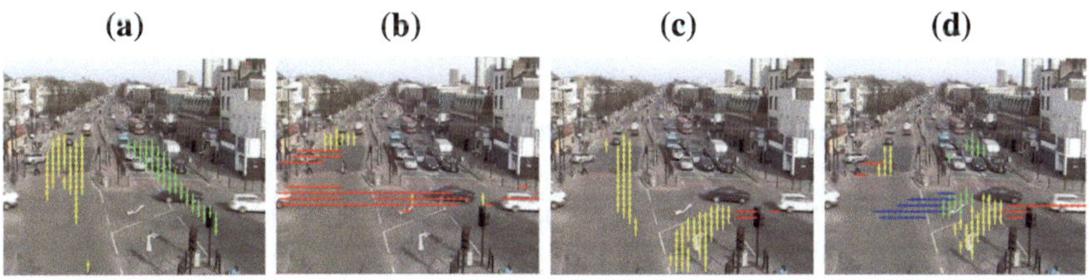

Fig. 4.5 Sample topics learnt by the dynamic HDP on the QMUL data. The arrows represent the visual words: the location and direction of the motion. **a** corresponds to the vertical traffic flow. **b** corresponds to the right traffic flow. **c** and **d** correspond to turns that follow the vertical traffic flow

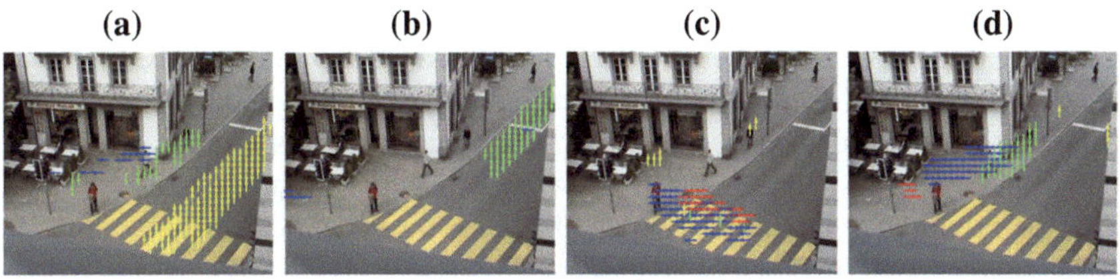

Fig. 4.6 Sample topics learnt by the dynamic HDP on the Idiap data. The arrows represent the visual words: the location and direction of the motion. **a** and **b** correspond to the upward and downward traffic flow, respectively. **c** and **d** correspond to pedestrian motions on a crosswalk and sidewalk, respectively

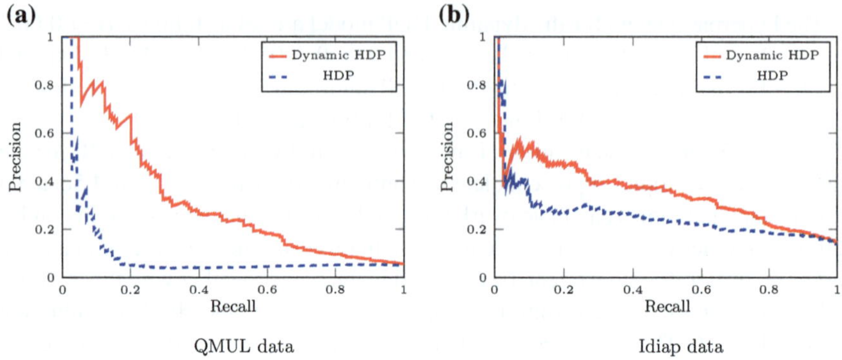

Fig. 4.7 Precision-recall curves obtained by the dynamic and vanilla HDP models on both datasets

Table 4.2 Mean area under precision-recall curves results for all topic models

Dataset	Dynamic HDP	EM	VB	GS
QMUL	0.3244	0.3166	0.3155	0.2970
Idiap	0.3565	0.3759	0.3729	0.3643

The proposed dynamic HDP is also compared with the algorithms from Chap. 3 on both datasets. For this 20 Monte Carlo runs are used with 5 posterior samples in every run. The obtained average results are presented in Table 4.2.

The dynamic HDP achieves the best results on the QMUL data. However, on the Idiap data MCTM methods demonstrate superior values of the performance measure. The Idiap data contains much more abnormal events: 11% of test documents are abnormal in comparison to 5% in the QMUL dataset. With these settings the definition of abnormal events as those that happen rarely is not as reasonable as for the QMUL data. Therefore, the flexibility of the dynamic HDP model can have a negative effect and abnormal activities might be learnt as typical. The limitation of model parameters prevents the MCTM methods from learning it.

To sum up, the proposed dynamic HDP algorithm outperforms the MCTM methods on the data with rare abnormal events. However, further developments are required to increase detection performance on data with considerable amounts of abnormal data points. Consideration of the dynamics is a promising direction here since the proposed dynamic HDP method is shown to significantly outperform its non-dynamic counterpart on both datasets.

4.6 Summary

In this chapter a novel Bayesian nonparametric dynamic topic model is proposed. The dynamics are considered on topic mixtures in documents such that successive documents are encouraged to have similar topic mixtures. This kind of dynamics for video processing is based on the fact that activities, expressed by topics, last for some time in real life. Therefore, activities presented in the current visual document are likely to be presented in the next document.

Note that the similar intuition is employed in the Markov clustering topic model considered in the previous chapter. The dynamic HDP introduced in this chapter represents a more flexible model in comparison to the MCTM due to its nonparametric nature. The concept of *behaviours*, namely the assumption of a limited number of topic mixtures, introduced in the MCTM is reasonable in the context of well-structured data where a general motion of an observed scene follows a cycle. Such type of data is analysed here; the data represents a video of a road junction regulated by a traffic light. The unlimitedness of the number of topics assumed in the dynamic HDP makes it applicable to more complex unstructured data, e.g., surveillance video in public places such as shopping malls or airports.

The inference algorithm for the proposed dynamic HDP is divided into two phases. On a training dataset the batch Gibbs sampler is applied. For making decision about test data the online Gibbs sampler is designed that allows to incrementally update the model without reprocessing previously observed data.

The proposed dynamic HDP topic model is applied for behaviour analysis and anomaly detection in video. A normality measure based on predictive likelihood is derived for decision making. Experiments on real video data show that the dynamic HDP significantly outperforms the conventional HDP in terms of anomaly detection performance. On the data with rare abnormal events the dynamic HDP also outperforms all the MCTM methods presented in the previous chapter.

Chapter 3 and this chapter present topic modeling methods for behaviour analysis and anomaly detection in video. An alternative change point detection approach is considered in the next chapter.

References

1. P. Orbanz, Y.W. Teh, Bayesian nonparametric models. in *Encyclopedia of Machine Learning* (Springer, Berlin, 2010), pp. 81–89
2. Y.W. Teh, M.I. Jordan, M.J. Beal, D.M. Blei, Hierarchical Dirichlet processes. J. Am. Stat. Assoc. **101**(476), 1566–1581 (2006)
3. T.S. Ferguson, A Bayesian analysis of some nonparametric problems. Ann. Stat. **1**(2), 209–230 (1973)
4. J. Sethuraman, A constructive definition of Dirichlet priors. Statistica sinica **4**(2), 639–650 (1994)

5. K. Vorontsov, O. Frei, M. Apishev, P. Romov, M. Dudarenko, BigARTM: open source library for regularized multimodal topic modeling of large collections. in *Proceedings of the 4th International Conference on Analysis of Images, Social Networks and Texts (AIST 2015)*, pp. 370–381, 2015
6. P. Smyth, M. Welling, A.U. Asuncion, Asynchronous distributed learning of topic models. in *Advances in Neural Information Processing Systems (NIPS 2009)*, pp. 81–88, 2009
7. C. Wang, J.W. Paisley, D.M. Blei, Online variational inference for the hierarchical Dirichlet process. in *Proceedings of the 14th International Conference on Artificial Intelligence and Statistics (AISTATS 2011)*, pp. 752–760, 2011
8. S.K. Pang, J. Li, S.J. Godsill, Detection and tracking of coordinated groups. IEEE Trans. Aerosp. Electron. Syst. **47**(1), 472–502 (2011)
9. T.L. Griffiths, M. Steyvers, Finding scientific topics. Proc. Natl. Acad. Sci. **101**(suppl 1), 5228–5235 (2004)

Chapter 5
Change Point Detection with Gaussian Processes

This chapter introduces a novel framework for detecting anomalies as change points. Chapters 3 and 4 present methods for behaviour analysis and anomaly detection based on the topic modeling approach. In this approach an algorithm extracts typical patterns as topics. During the testing phase, data which does not fit with the extracted patterns, is labelled as abnormal. Anomaly detection can be viewed from another perspective. Anomalies can be considered as a change in an underlying data distribution, employing the change point detection methodology for anomaly discovery. For example, this approach is relevant in such classes of applications as panic or evacuation detection, where input data can be an average velocity of a crowd.

This chapter presents a general approach for change point detection, which can be used for behaviour analysis (where periods between change points are considered as different behaviours) and anomaly detection (where a change is considered as a break point between normal and abnormal behaviours). In the proposed framework changes are considered as functional breaks in input data. Functions governing observed data have a Gaussian process prior and a change is defined as an alteration in hyperparameters of this Gaussian process. A hypothesis testing framework is employed and statistical tests for change point detection are proposed.

In this chapter input data is presented in a time series form. An overview of the methods that process video data in such form is provided in Sect. 2.4.2.

The results of the work presented in this chapter are disseminated in:

- O. Isupova, D. Kuzin, F. Gustafsson, L. Mihaylova. "Change Point Detection with Gaussian Processes", in *IEEE Transactions on Pattern Analysis and Machine Intelligence*, under review, 2017

The rest of the chapter is organised as follows. Section 5.1 formulates a change point detection problem for a Gaussian process time series model within a statistical hypothesis testing framework. Section 5.2 describes the proposed likelihood ratio tests. Online statistical tests are introduced in Sect. 5.3. The proposed methods are

O. Isupova, *Machine Learning Methods for Behaviour Analysis and Anomaly Detection in Video*, Springer Theses, https://doi.org/10.1007/978-3-319-75508-3_5

evaluated on simulated data in Sect. 5.4 and on real data in Sect. 5.5. Section 5.6 summarises the chapter.

5.1 Problem Formulation

Change point detection aims to detect abrupt changes in time series data. An abrupt change is understood as a change in a latent probability distribution of observed data.

5.1.1 Data Model

A time series can be modelled with a Gaussian process (GP) as follows:

$$y_\tau = f(\tau) + \varepsilon_\tau, \tag{5.1}$$

where y_τ is an observation at time τ; $f(\cdot) \sim \mathcal{GP}(m(\cdot), k(\cdot, \cdot))$ is a function of time with a GP prior, where $\mathcal{GP}$ denotes a GP, characterised by a mean function $m(\cdot)$ and a covariance function $k(\cdot, \cdot)$; $\varepsilon_\tau \sim \mathcal{N}(\varepsilon_\tau|0, \sigma^2)$ is white Gaussian noise with a zero mean and a variance σ^2.

The mean and covariance functions of the GP are parameterised with vectors $\boldsymbol{\vartheta}^m$ and $\boldsymbol{\vartheta}^k$, respectively. Then $\boldsymbol{\vartheta} = \{\boldsymbol{\vartheta}^m, \boldsymbol{\vartheta}^k, \sigma^2\}$ denotes a vector of hyperparameters of a given GP.

Although in this thesis we consider only the GP time series model, the proposed methods are directly applicable to the GP autoregressive model:

$$y_\tau = f(y_{\tau-r}, \ldots, y_{\tau-1}) + \varepsilon_\tau,$$

where r is the order of the model.

For a given set of input points of time indices $\boldsymbol{\tau}_{1:N} = \{\tau_i\}_{i=1}^N$, where N is the number of points, we can compute the posterior distributions of function values and observations [1]:

$$\mathbf{f}_{1:N}|\boldsymbol{\tau}_{1:N}, \boldsymbol{\vartheta} \sim \mathcal{N}(\mathbf{f}_{1:N}|\boldsymbol{\mu}, \mathbf{K}), \tag{5.2}$$

where $\mathbf{f}_{1:N} = f(\boldsymbol{\tau}_{1:N}) = \{f(\tau_i)\}_{i=1}^N$ are function f values for the given input time indices; $\boldsymbol{\mu} = \{\mu_i\}_{i=1}^N = \{m(\tau_i)\}_{i=1}^N$ are realisations of the GP mean function at the input points; $\mathbf{K} = \{\mathbf{K}_{i,j}\}_{i,j=1}^N = \{k(\tau_i, \tau_j)\}_{i,j=1}^N$ are realisations of the GP covariance function at the input points; and:

$$\mathbf{y}_{1:N}|\boldsymbol{\tau}_{1:N}, \boldsymbol{\vartheta} \sim \mathcal{N}(\mathbf{y}_{1:N}|\boldsymbol{\mu}, \mathbf{K} + \sigma^2\mathbf{I}), \tag{5.3}$$

where $\mathbf{y}_{1:N} = \{y_{\tau_i}\}_{i=1}^N$ are observations at the given time indices and $\mathbf{I}$ is the identity matrix.

The marginal log likelihood function of observed data is given by:

$$\log p(\mathbf{y}_{1:N}|\boldsymbol{\tau}_{1:N}, \boldsymbol{\vartheta}) = -\frac{1}{2}\left(\mathbf{y}_{1:N} - \boldsymbol{\mu}\right)^T \left(\mathbf{K} + \sigma^2\mathbf{I}\right)^{-1} \left(\mathbf{y}_{1:N} - \boldsymbol{\mu}\right) - \frac{1}{2}\log\det\left(\mathbf{K} + \sigma^2\mathbf{I}\right) - \frac{N}{2}\log 2\pi, \tag{5.4}$$

where $\det(\cdot)$ is a determinant of a matrix.

5.1.2 Change Point Detection Problem Formulation

An abrupt change in GP time series data can be defined as a change in hyperparameters of the GP. This means that the function $f(\cdot)$ from the data model (5.1) has a GP prior governed by different hyperparameters before and after a change. The vectors of hyperparameters before and after a change are denoted as $\boldsymbol{\vartheta}_0$ and $\boldsymbol{\vartheta}_1$, respectively. The final model for time series data with an abrupt change is then formulated as:

$$y_\tau = f_0(\tau) + \varepsilon_\tau^0, \qquad \forall \tau < \tau^*; \tag{5.5}$$

$$y_\tau = f_1(\tau) + \varepsilon_\tau^1, \qquad \forall \tau \geq \tau^*, \tag{5.6}$$

where τ^* is a time instant when the change occurs and:

$$f_0(\cdot) \sim \mathcal{GP}(m_0(\cdot), k_0(\cdot, \cdot)); \tag{5.7a}$$

$$\varepsilon_\tau^0 \sim \mathcal{N}(0, \sigma_0^2); \tag{5.7b}$$

$$f_1(\cdot) \sim \mathcal{GP}(m_1(\cdot), k_1(\cdot, \cdot)); \tag{5.7c}$$

$$\varepsilon_\tau^1 \sim \mathcal{N}(0, \sigma_1^2), \tag{5.7d}$$

where m_a and k_a are the mean and covariance functions of the GP governed by the hyperparameter vector $\boldsymbol{\vartheta}_a$, σ_a^2 is the variance of the additive noise defined by the hyperparameter vector $\boldsymbol{\vartheta}_a$, for $a \in \{0, 1\}$.

The change point detection problem can be formulated within a statistical framework as a hypothesis testing task. Let the null hypothesis $\mathcal{H}_0$ state that a GP prior of a function f has a hyperparameter vector $\boldsymbol{\vartheta}$ equal to the initial value $\boldsymbol{\vartheta}_0$ during the whole observation period. Meanwhile let the alternative $\mathcal{H}_1$ claim that there exists a change time index τ^* such that the hyperparameter vector $\boldsymbol{\vartheta}$ of the GP has the initial value $\boldsymbol{\vartheta}_0$ before the change point and is equal to a new vector $\boldsymbol{\vartheta}_1$ after the change time τ^*:

$$\mathcal{H}_0: \qquad \boldsymbol{\vartheta} = \boldsymbol{\vartheta}_0, \quad \forall \tau; \tag{5.8}$$

$$\mathcal{H}_1: \exists \tau^*: \begin{cases} \boldsymbol{\vartheta} = \boldsymbol{\vartheta}_0, & \tau < \tau^*, \\ \boldsymbol{\vartheta} = \boldsymbol{\vartheta}_1, & \tau \geq \tau^*. \end{cases} \tag{5.9}$$

The next sections provide a description of the proposed statistical methods for the testing framework introduced in (5.8)–(5.9).

5.2 Gaussian Process Change Point Detection Approach Based on Likelihood Ratio Tests

This section introduces an approach based on two likelihood ratio tests for the change point detection problem (5.8)–(5.9). The first proposed test is the likelihood ratio test (LRT) for the considered problem. In this set-up both hyperparameter values $\boldsymbol{\vartheta}_0$ and $\boldsymbol{\vartheta}_1$ are assumed to be known. In some class of the real world applications this assumption is unlikely to hold therefore the second proposed test is the generalised LRT, where the hyperparameter vectors $\boldsymbol{\vartheta}_0$ and $\boldsymbol{\vartheta}_1$ are estimated from the data.

5.2.1 *Likelihood Ratio Test*

Let $\mathbf{y}_{1:N}$ be observed output variables, $\boldsymbol{\vartheta}_0$ and $\boldsymbol{\vartheta}_1$ are known hyperparameter vectors of the GP before and after a change, respectively. Define a log likelihood ratio test statistic for change point detection at time τ^* as:

$$T_{\text{LRT}}(\tau^*|\mathbf{y}_{1:N}) \stackrel{\text{def}}{=} 2\log\left(\frac{p(\mathbf{y}_{1:\tau^*}|\boldsymbol{\tau}_{1:\tau^*}, \boldsymbol{\vartheta}_0)p(\mathbf{y}_{\tau^*+1:N}|\boldsymbol{\tau}_{\tau^*+1:N}, \boldsymbol{\vartheta}_1)}{p(\mathbf{y}_{1:N}|\boldsymbol{\tau}_{1:N}, \boldsymbol{\vartheta}_0)}\right) =$$
$$2\log p(\mathbf{y}_{1:\tau^*}|\boldsymbol{\tau}_{1:\tau^*}, \boldsymbol{\vartheta}_0) + 2\log p(\mathbf{y}_{\tau^*+1:N}|\boldsymbol{\tau}_{\tau^*+1:N}, \boldsymbol{\vartheta}_1) - 2\log p(\mathbf{y}_{1:N}|\boldsymbol{\tau}_{1:N}, \boldsymbol{\vartheta}_0) \tag{5.10}$$

where $\log p(\mathbf{y}_{1:\tau^*}|\boldsymbol{\tau}_{1:\tau^*}, \boldsymbol{\vartheta}_0)$ is the log likelihood of the data starting from the first time moment till the time moment τ^* computed based on the GP hyperparameter vector $\boldsymbol{\vartheta}_0$; $\log p(\mathbf{y}_{\tau^*+1:N}|\boldsymbol{\tau}_{\tau^*+1:N}, \boldsymbol{\vartheta}_1)$ is the likelihood of the data starting after the time τ^* till the end of the observation period given the hyperparameter vector $\boldsymbol{\vartheta}_1$; $\log p(\mathbf{y}_{1:N}|\boldsymbol{\tau}_{1:N}, \boldsymbol{\vartheta}_0)$ is the likelihood of the whole set of observations given the hyperparameter vector $\boldsymbol{\vartheta}_0$.

Substituting the log likelihood expression (5.4) for the GP data the test statistic is given as:

$$T_{\mathrm{LRT}}(\tau^*|\mathbf{y}_{1:N}) = -\left(\tilde{\mathbf{y}}_0^{1:\tau^*}\right)^T \left(\tilde{\mathbf{K}}_0^{1:\tau^*}\right)^{-1} \tilde{\mathbf{y}}_0^{1:\tau^*} - \log\det\left(\tilde{\mathbf{K}}_0^{1:\tau^*}\right) - \left(\tilde{\mathbf{y}}_1^{\tau^*+1:N}\right)^T \left(\tilde{\mathbf{K}}_1^{\tau^*+1:N}\right)^{-1} \tilde{\mathbf{y}}_1^{\tau^*+1:N} - \log\det\left(\tilde{\mathbf{K}}_1^{\tau^*+1:N}\right) + \left(\tilde{\mathbf{y}}_0^{1:N}\right)^T \left(\tilde{\mathbf{K}}_1^{1:N}\right)^{-1} \tilde{\mathbf{y}}_0^{1:N} + \log\det\left(\tilde{\mathbf{K}}_1^{1:N}\right), \tag{5.11}$$

where $\tilde{\mathbf{y}}_a = \mathbf{y} - \boldsymbol{\mu}_a$ are the centralised observations and $\tilde{\mathbf{K}}_a = \mathbf{K}_a + \sigma_a^2\mathbf{I}$ is the covariance function of the observed data, both the mean vector and covariance matrix are computed given the hyperparameter vector $\boldsymbol{\vartheta}_a$, for $a \in \{0, 1\}$; superscript denotes the input point indices, for which the corresponding variable is obtained.

The statistic $T_{\mathrm{LRT}}(\tau^*|\mathbf{y}_{1:N})$ represents the log likelihood of the data under assumption that a change occurs at time τ^* in proportion of the likelihood of the data without a change.

The likelihood ratio test to estimate the change time τ^* can be formulated in the following way.

Definition 1 **The likelihood ratio change point detection test** is defined as:

$$\max_{\tau^*} T_{\mathrm{LRT}}(\tau^*|\mathbf{y}_{1:N}) > c_{\mathrm{LRT,THR}}, \tag{5.12}$$

where T_{LRT} is given by (5.10) and $c_{\mathrm{LRT,THR}}$ is a threshold. If the threshold $c_{\mathrm{LRT,THR}}$ is exceeded, the time moment τ^* that maximises T_{LRT} is chosen as the estimated change time.

For any given τ^*, the LRT is the optimal test according to the Neyman-Pearson lemma [2].

The threshold $c_{\mathrm{LRT,THR}}$ value should be greater than 0, as the statistic $T_{\mathrm{LRT}}(N|\mathbf{y}_{1:N})$ value of the last time moment is equal to zero and this case by convention corresponds to the "no change" decision by the test.

5.2.2 *Generalised Likelihood Ratio Test*

The proposed generalised LRT allows to relax the assumption that the hyperparameters of the GP are known. The generalised LRT is used as LRT (Definition 1), but values for the vectors $\boldsymbol{\vartheta}_0$ and $\boldsymbol{\vartheta}_1$ are estimated from the data. The generalised LRT statistic is given by:

$$T_{\mathrm{gLRT}}(\tau^*|\mathbf{y}_{1:N}) \stackrel{\mathrm{def}}{=} 2\log\left(\frac{p(\mathbf{y}_{1:\tau^*}|\boldsymbol{\tau}_{1:\tau^*},\hat{\boldsymbol{\vartheta}}_0)\,p(\mathbf{y}_{\tau^*+1:N}|\boldsymbol{\tau}_{\tau^*+1:N},\hat{\boldsymbol{\vartheta}}_1)}{p(\mathbf{y}_{1:N}|\boldsymbol{\tau}_{1:N},\hat{\boldsymbol{\vartheta}}_0)}\right) =$$
$$2\log p(\mathbf{y}_{1:\tau^*}|\boldsymbol{\tau}_{1:\tau^*},\hat{\boldsymbol{\vartheta}}_0) + 2\log p(\mathbf{y}_{\tau^*+1:N}|\boldsymbol{\tau}_{\tau^*+1:N},\hat{\boldsymbol{\vartheta}}_1) - 2\log p(\mathbf{y}_{1:N}|\boldsymbol{\tau}_{1:N},\hat{\boldsymbol{\vartheta}}_0), \quad (5.13)$$

where $\hat{\boldsymbol{\vartheta}}_0$ and $\hat{\boldsymbol{\vartheta}}_1$ are estimates of the hyperparameter vectors of the GP.

The estimates of the hyperparameters can be obtained by maximising the marginal likelihood (5.4):

$$\hat{\boldsymbol{\vartheta}}_a = \underset{\boldsymbol{\vartheta}_a}{\operatorname{argmax}}\, \log p(\mathbf{y}|\boldsymbol{\tau},\boldsymbol{\vartheta}_a) =$$
$$\underset{\boldsymbol{\vartheta}_a}{\operatorname{argmax}}\left(-\frac{1}{2}\left(\mathbf{y}-\boldsymbol{\mu}_a\right)^T\left(\mathbf{K}_a+\sigma_a^2\mathbf{I}\right)^{-1}\left(\mathbf{y}-\boldsymbol{\mu}_a\right)-\right.$$
$$\left.\frac{1}{2}\log\det\left(\mathbf{K}_a+\sigma_a^2\mathbf{I}\right)-\frac{N}{2}\log 2\pi\right), \quad a\in\{0,1\}. \quad (5.14)$$

Details of the GP hyperparameter optimisation can be found in Appendix E.

5.2.3 *Discussion*

The likelihood ratio tests represent the core elements in change point detection and realisations of these tests for different problems are widely used in the literature [2]. These tests are simple to interpret; they are easily implemented. The likelihood ratio test with known hyperparameters is proven to be optimal. The generalised likelihood ratio test uses the estimates of the hyperparameters and it might be expected to be near optimal if the estimates of the hyperparameters are close enough to the true unknown values.

However, in real-world applications a practitioner might face some issues using the likelihood ratio tests. Firstly, tests are designed for offline data processing. This means the whole dataset is required to start running the tests. It can be inappropriate for settings when a decision should be made in an online manner and some actions are expected in the case of an alarm about a change. For example, a change can mean some error in an industrial line production and the line should be fixed as soon as possible in order to reduce the number of defected products.

Secondly, the tests expect no more than one change point. Even if online processing is not essential, this limitation to have no more than one change point implies further constraints on applicability of the tests.

In order to overcome these issues another family of statistical tests is proposed in the next section. These tests work within a sliding window allowing both online data processing and multiple change points detection.

5.3 Gaussian Process Online Change Point Detection Approach Based on Likelihood Estimation

This section introduces an online approach that relies on statistical tests that process sequential upcoming data. Let $\mathbf{y}_{1:\tau}$ be observed data up to the current moment τ, followed the model (5.5)–(5.6). At each time τ we are interested if there is a change in a hyperparameter vector $\boldsymbol{\vartheta}$ from its initial value $\boldsymbol{\vartheta}_0$. Note that the new value $\boldsymbol{\vartheta}_1$ is neither known nor estimated from the data to make a decision about a change. After a change has been detected, it may be relevant to estimate the new set of hyperparameters, and then restart the test to look for the next change.

5.3.1 Test Formulation

The following test statistic is computed for data within a sliding window of length L:

$$T_{\text{online}}(\mathbf{y}_{\tau-L+1:\tau}) \stackrel{\text{def}}{=} -2\log p(\mathbf{y}_{\tau-L+1:\tau}|\boldsymbol{\tau}_{1:\tau}, \boldsymbol{\vartheta}_0) = -2\log p(\mathbf{y}_{\tau-L+1:\tau}|\boldsymbol{\tau}_{\tau-L+1:\tau}, \boldsymbol{\vartheta}_0) =$$
$$(\mathbf{y}_{\tau-L+1:\tau} - \boldsymbol{\mu}_0)^T \left(\mathbf{K}_0 + \sigma_0^2\mathbf{I}\right)^{-1} (\mathbf{y}_{\tau-L+1:\tau} - \boldsymbol{\mu}_0) + \log\det\left(\mathbf{K}_0 + \sigma_0^2\mathbf{I}\right) - L\log 2\pi \tag{5.15}$$

Theorem 1 *Consider the problem (5.8)–(5.9), where $\mathbf{y}_{1:\tau}$ is the observed data up to time τ and $\boldsymbol{\vartheta}_0$ is a known hyperparameter vector of the GP before a change. Let $T_{online}(\mathbf{y}_{\tau-L+1:\tau})$ be the likelihood-based statistic defined as (5.15). Then given that the null hypothesis $\mathcal{H}_0$ is true:*

$$T_{online}(\mathbf{y}_{\tau-L+1:\tau}) - \log\det\left(\mathbf{K}_0 + \sigma_0^2\mathbf{I}\right) - L\log 2\pi \sim \chi_L^2, \tag{5.16}$$

where χ_L^2 is a chi-squared distribution with L degrees of freedom.

Proof of the Theorem 1 is given in Appendix D.

Figure 5.1 shows a visual comparison of an empirical normalised histogram of the statistic $T_{\text{online}} - \log\det\left(\mathbf{K}_0 + \sigma_0^2\mathbf{I}\right) - L\log 2\pi$ and the analytical χ_L^2 distribution. The normalised histogram is built based on 10 000 Monte Carlo simulations with $\mathbf{y} \sim \mathcal{N}\left(\mathbf{y}|\boldsymbol{\mu}_0 = \mathbf{0}, \mathbf{K}_0\right)$, where $\mathbf{K}_0$ is a squared exponential covariance matrix:

$$\mathbf{K}_0(i, j) = \sigma_{\mathbf{K}}^2 \exp\left(-\frac{(i-j)^2}{2\ell^2}\right) \tag{5.17}$$

with the signal variance $\sigma_{\mathbf{K}}^2 = 0.5$ and the length scale $\ell = 1$. The length of the vector $\mathbf{y}$ is 10. Noise-free observations are considered, therefore $\sigma_0^2 = 0$. The Kolmogorov-Smirnov test does not reject a hypothesis that the sample of T_{online} is obtained from the corresponding χ_{10}^2 distribution with a significance level 0.05 and p-value of 0.495.

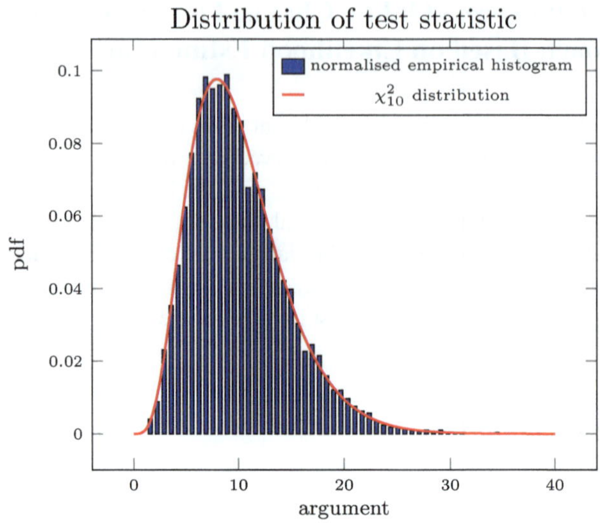

Fig. 5.1 Empirical normalised histogram of the test likelihood-based statistic with the corresponding analytical χ^2 distribution

Once the statistic distribution under the null hypothesis is determined the test procedure can be defined.

Definition 2 **The online likelihood-based change point detection test** is defined as:

Define $T_{\text{online}}(\mathbf{y}_{\mathcal{T}-L+1:\mathcal{T}})$ as given in (5.15), which is χ^2_L-distributed under the null hypothesis $\mathcal{H}_0$. Let e_1 and e_2 be quantiles of the χ^2_L distribution such that: $\mathcal{F}_{\chi^2_L}(e_1) = \frac{\alpha_{\text{stat}}}{2}$ and $\mathcal{F}_{\chi^2_L}(e_2) = 1 - \frac{\alpha_{\text{stat}}}{2}$, where $\mathcal{F}_{\chi^2_L}(\cdot)$ is the cumulative density function (cdf) of the χ^2_L distribution and α_{stat} is a given significance level.

Reject $\mathcal{H}_0$ if $T_{\text{online}}(\mathbf{y}_{\mathcal{T}-L+1:\mathcal{T}}) < e_1$ or $T_{\text{online}}(\mathbf{y}_{\mathcal{T}-L+1:\mathcal{T}}) > e_2$.

5.3.2 Theoretical Evaluation of the Test

Effectiveness of a statistical test is usually evaluated by probabilities of type-I and type-II errors. A type-I error is a rejection of the null hypothesis when it is true (i.e., a false alarm), while a type-II is a failure to reject the null hypothesis while the alternative is true (i.e., a missed detection). The probability of the type-I error for the proposed test is fixed and equal to α_{stat}. We do not evaluate the probability β_{stat} of the type-II error directly, we rather evaluate the power of the test, which is a complement to β_{stat}.

The power of the test is defined as [3]:

$$B(\vartheta) \overset{\text{def}}{=} \mathbb{P}(\text{reject } \mathcal{H}_0 | \mathcal{H}_1 \text{is true}). \tag{5.18}$$

The following distribution is useful in the context of the proposed test and its power:

Definition 3 Let ζ_i be independent random variables having a non-central chi-squared distribution and a_i be some coefficients, $i \in \{1, \dots, I\}$. Then a random variable $\tilde{\zeta}$:

$$\tilde{\zeta} = \sum_{i=1}^{I} a_i \zeta_i \tag{5.19}$$

has a **generalised chi-squared distribution.**

A generalised chi-squared distribution has no closed-form expression but can be efficiently estimated numerically [4].

Theorem 2 *Consider the problem (5.8)–(5.9) and the online likelihood-based change point detection test defined in Definition 2. Then the power of this test is:*

$$B(\vartheta) = 1 + \mathcal{F}_\beta(e_1) - \mathcal{F}_\beta(e_2), \tag{5.20}$$

where $\mathcal{F}_\beta(\cdot)$ is the cdf of the random variable β. The variable β follows the generalised chi-squared distribution plus a displacement:

$$\beta = \sum_{i=1}^{L} d_i v_i + \log\det\left(\mathbf{K}_0 + \sigma_0^2 \mathbf{I}\right) + L \log 2\pi, \tag{5.21}$$

where d_i are eigenvalues of the matrix **A***:*

$$\mathbf{A} = \left(\mathbf{K}_1 + \sigma_1^2 \mathbf{I}\right)^{\frac{1}{2}} \left(\mathbf{K}_0 + \sigma_0^2 \mathbf{I}\right)^{-1} \left(\mathbf{K}_1 + \sigma_1^2 \mathbf{I}\right)^{\frac{1}{2}}, \tag{5.22}$$

and v_i are random variables: $v_i \sim \chi_1'^2(o_i^2)$ while o_i are components of a vector $\mathbf{o} = \mathbf{P}^T \left(\mathbf{K}_1 + \sigma_1^2 \mathbf{I}\right)^{-\frac{1}{2}} \left(\boldsymbol{\mu}_1 - \boldsymbol{\mu}_0\right)$, where **P** *is a matrix, where columns are eigenvectors of the matrix* **A**.

Proof of the Theorem 2 is presented in Appendix D.

Theorem 2 allows calculating of the theoretical success rate of test application for any given data distribution hyperparameters ϑ_1.

The sliding window nature of the proposed test (Definition 2) makes it applicable for online data processing. In practice the true hyperparameter vector ϑ_0, which can be unknown, may be replaced by its estimate. Note that the data that is used for hyperparameter vector estimation is not limited to the sliding window employed in the test statistic computation. More historical data can be used, therefore an accurate estimate might be expected. Moreover, if the estimates of the hyperparameters are updated based on recent data the test would be able to detect multiple changes between different data distributions.

5.3.3 Test with Estimated Hyperparameters

Consider the online likelihood-based change point detection test (Definition 2) when the initial hyperparameter vector $\boldsymbol{\vartheta}_0$ is unknown. The value of $\boldsymbol{\vartheta}_0$ can be estimated using previously observed data $\mathbf{y}_{\tau':\tau''}$, where $\tau' \leq \tau'' < \tau$, for example, by maximising the marginal likelihood:

$$\hat{\boldsymbol{\vartheta}}_0 = \underset{\boldsymbol{\vartheta}_0}{\operatorname{argmax}} \log p(\mathbf{y}_{\tau':\tau''} | \boldsymbol{\tau}_{\tau':\tau''}, \boldsymbol{\vartheta}_0). \tag{5.23}$$

The test statistic $T_{\text{online est}}$ can then be computed as:

$$T_{\text{online est}}(\mathbf{y}_{\tau-L+1:\tau}) \stackrel{\text{def}}{=} -2 \log p(\mathbf{y}_{\tau-L+1:\tau} | \boldsymbol{\tau}_{1:\tau}) \approx -2 \log p(\mathbf{y}_{\tau-L+1:\tau} | \boldsymbol{\tau}_{\tau-L+1:\tau}, \hat{\boldsymbol{\vartheta}}_0). \tag{5.24}$$

Here τ' and τ'' are set to form another sliding window of length $\tilde{L}$: $\tau' = \tau - \tilde{L}$ and $\tau'' = \tau - 1$. In such settings the estimate $\hat{\boldsymbol{\vartheta}}_0$ of the hyperparameter vector is always updated based on the most recent data and the test can be applied in an online manner. Moreover, it allows to adapt the test to different data distributions and the test is able to detect changes between the periods with stationary data distributions, i.e., when the data model is:

$$y_\tau = f_0(\tau) + \varepsilon_\tau^0, \qquad \forall \tau < \tau_1^*, \qquad \boldsymbol{\vartheta} = \boldsymbol{\vartheta}_0; \tag{5.25}$$

$$y_\tau = f_1(\tau) + \varepsilon_\tau^1, \qquad \forall \tau_1^* \leq \tau < \tau_2^*, \qquad \boldsymbol{\vartheta} = \boldsymbol{\vartheta}_1; \tag{5.26}$$

$$y_\tau = f_2(\tau) + \varepsilon_\tau^2, \qquad \forall \tau_2^* \leq \tau < \tau_3^*, \qquad \boldsymbol{\vartheta} = \boldsymbol{\vartheta}_2; \tag{5.27}$$

$$\ldots$$

and the goal is to detect all change points $\tau_1^*, \tau_2^*, \tau_3^*, \ldots$ Due to the sliding window approach the test after the change point detection adapts to new data allowing detection of a new change point breaking the new data stationary regime.

5.3.4 Discussion

The tests proposed in this section provide a tool for online change point detection. As a GP represents a prior for a wide variety of functions the proposed approach is general and can be applied in very different scenarios. For example, when data is expected to follow a smooth function a squared exponential covariance function can be used. If data is affected by seasonal variations, periodic covariance functions can come into play.

To use the proposed tests, the particular form of a GP specified by a choice of a hyperparameter vector is required only for data before a change. No assumptions about the distribution of data after a change are necessary. If data starts to follow a new rule, e.g., a different covariance function is used in a GP, the method will detect a change.

5.4 Performance Validation on Synthetic Data

This section presents a numerical evaluation of the proposed methods for change point detection on synthetic data.

The following performance measures are used [2, 3]:

- *Mean time between false alarms*: $\mathbb{E}(\tau_a - \tau_0 |$ no change$)$, where τ_a is a time of an alarm, generated by a change point detection algorithm, and this alarm is false, τ_0 is a starting time. We consider the previous false alarm or a true missed change point as a starting time;
- *Mean delay for detection*: $\mathbb{E}(\tau_a - \tau^*)$, where τ^* is a true time of a change;
- *Missed detection rate*: a probability of not receiving an alarm, when there has been a change, which can be estimated as $\frac{\#\{\text{missed detections}\}}{\#\{\text{change points}\}}$.

A change point detection algorithm is required to maximise the first measure and minimise the other two. A reliable algorithm should achieve a tradeoff with respect to the three criteria. Increase of the number of algorithm detections leads to reduction of the missed detection rate, although there is a risk to increase the number of false alarms and hence it might negatively affect the mean time between false alarms measure. On the other hand, reduction of the number of detections might improve the performance in terms of false alarms, however, it might also increase the missed detection rate.

It is a well-known problem [2, 3] that it is unclear how to distinguish correct detections and false alarms. For example, an alarm at time point $\tau_a > \tau^*$, where τ^* is a true time of a change, can be considered as a delayed correct detection or as a false alarm. In this work when the detection comes within a time window around the true change point it is considered as a correct detection. The time difference between this detection and the true change point contributes to the mean delay for detection performance measure. If the detection comes outside of this time window it is treated as a false alarm. Therefore, the mean time between false alarms is updated.

5.4.1 Data Simulated by the Proposed Generative Model

A time series generated by the model (5.5)–(5.6) is considered, where the change point time τ^* is set to 101. The total number N of observed points is equal to 200. Observed data points are acquired at time moments $\boldsymbol{\tau}_{1:N} = 1, 2, \ldots, 200$. Function values are sampled based on the GP: $\mathbf{f}_{1:100} \sim \mathcal{N}(\mathbf{f}_{1:100}|\mathbf{0}, \mathbf{K}_0)$ and $\mathbf{f}_{101:200} \sim \mathcal{N}(\mathbf{f}_{101:200}|\mathbf{0}, \mathbf{K}_1)$ before and after the change, respectively. Squared exponential covariance matrices are used:

$$\mathbf{K}_a(i, j) = \sigma^2_{\mathbf{K}_a} \exp\left(-\frac{(\tau_i - \tau_j)^2}{2\ell_a^2}\right), \quad a \in \{0, 1\}, \tag{5.28}$$

which have two parameters: a signal variance $\sigma^2_{\mathbf{K}_a}$ and a length scale ℓ_a.

Corresponding noisy observations $\mathbf{y}_{1:100}$ and $\mathbf{y}_{101:200}$ are obtained by adding the Gaussian-distributed random noise, which variance is set to $\sigma^2 = 0.1$ both before and after the change.

Theorem 2 allows to compute the power of the GP Online Likelihood-based Change point Detection Test (GP-OLCDT) for any given hyperparameter vectors $\boldsymbol{\vartheta}_0$ and $\boldsymbol{\vartheta}_1$. In the considered settings only the hyperparameters of a covariance function are changing: $\boldsymbol{\vartheta}_a = \{\sigma^2_{\mathbf{K}_a}, \ell_a\}, a \in \{0, 1\}$.

Figure 5.2 presents the power of the test values as a function of the hyperparameter vector $\boldsymbol{\vartheta}_1$ for the alternative $\mathcal{H}_1$. The hyperparameter vector $\boldsymbol{\vartheta}_0$ for the null hypothesis $\mathcal{H}_0$ is fixed and its components are assigned as: $\ell_0 = 3$ and $\sigma^2_{\mathbf{K}_0} = 1$. In the current settings the proposed test performs well in the areas where the length scale is less after the change than before it and the signal variance changes from small to large values before and after the change, respectively.

The following scenarios are examined in more detail:

1. $\ell_0 > \ell_1$ and $\sigma^2_{\mathbf{K}_0} = \sigma^2_{\mathbf{K}_1}$;
2. $\ell_0 < \ell_1$ and $\sigma^2_{\mathbf{K}_0} = \sigma^2_{\mathbf{K}_1}$;
3. $\ell_0 = \ell_1$ and $\sigma^2_{\mathbf{K}_0} > \sigma^2_{\mathbf{K}_1}$;
4. $\ell_0 = \ell_1$ and $\sigma^2_{\mathbf{K}_0} < \sigma^2_{\mathbf{K}_1}$.

Based on the results given in Fig. 5.2 the proposed GP-OLCDT is expected to perform better under the 1-st and 4-th scenarios while having difficulties under the 2-nd and 3-rd scenarios. Indeed, in the 2-nd and 3-rd scenarios the hyperparameter vector $\boldsymbol{\vartheta}_0$ value before a change represents more flexible settings that allow to explain larger variety of data. Therefore, the data after the change, based on the hyperparameter vector $\boldsymbol{\vartheta}_1$, can be also fitted with the model before the change. The likelihood function value would not then drop significantly after the change, which causes low performance of the test.

The distribution of the test statistic for the GP-OLCDT can be determined based on Theorem 1. The cdfs of the test statistic under the null hypothesis $\mathcal{H}_0$ and alternative $\mathcal{H}_1$ for all four scenarios are presented in Fig. 5.3. The figure demonstrates that the cdfs of the test statistic under the null hypothesis and alternative are very close to

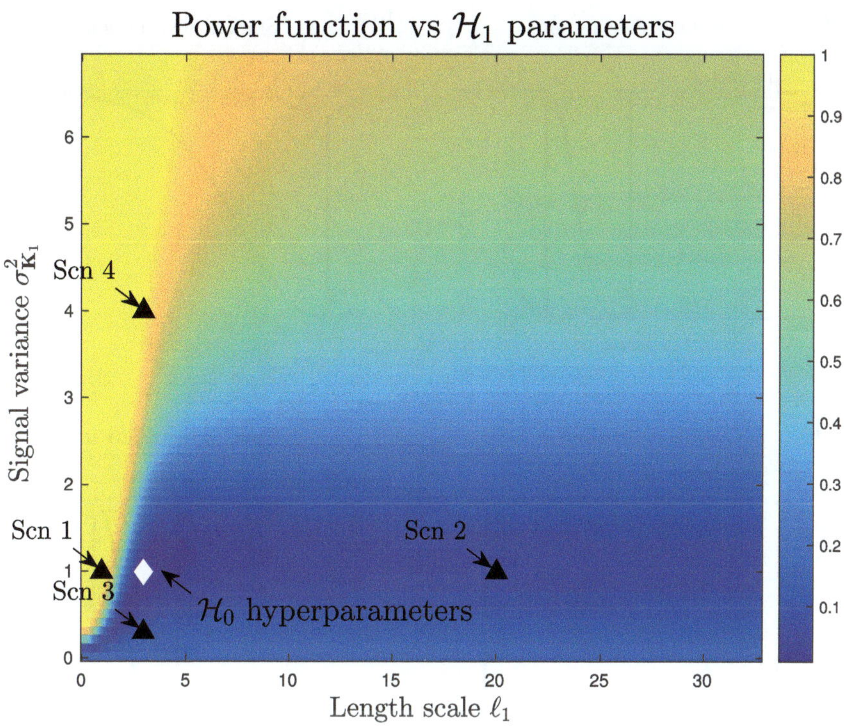

Fig. 5.2 Power of the GP-OLCDT as a function of the hyperparameters for the alternative $\mathcal{H}_1$. The null hypothesis hyperparameters are plotted as a diamond. Triangular points correspond to the values of the alternative hyperparameters from the four considered scenarios

each other for the 2-nd and 3-rd scenarios and it is difficult to distinguish the two cdfs. It further explains the reasons of low power of the test values for these two scenarios.

For each of the scenarios the following methods are applied:

- the GP-OLCDT with known ϑ_0 (Sect. 5.3.1);
- the GP-OLCDT with batch estimated $\hat{\vartheta}_0$, where the estimate is obtained by optimising marginal likelihood on the first part of data $\mathbf{y}_{1:100}$ (Sect. 5.3.1);
- the GP-OLCDT with sliding estimated $\hat{\vartheta}_0$, where the estimate is obtained by optimising marginal likelihood on data $\mathbf{y}_{\tau-\tilde{L}:\tau-1}$ from a sliding window (Sect. 5.3.3), $\tilde{L} = 25$;
- the GP log likelihood ratio change point detection test (GP-lLRT) with known ϑ_0 and ϑ_1 (Sect. 5.2.1);
- the GP generalised log likelihood ratio change point detection test (GP-glLRT) with estimated $\hat{\vartheta}_0$ and $\hat{\vartheta}_1$, where the estimates are obtained based on first and last 25 data points, respectively (Sect. 5.2.2).

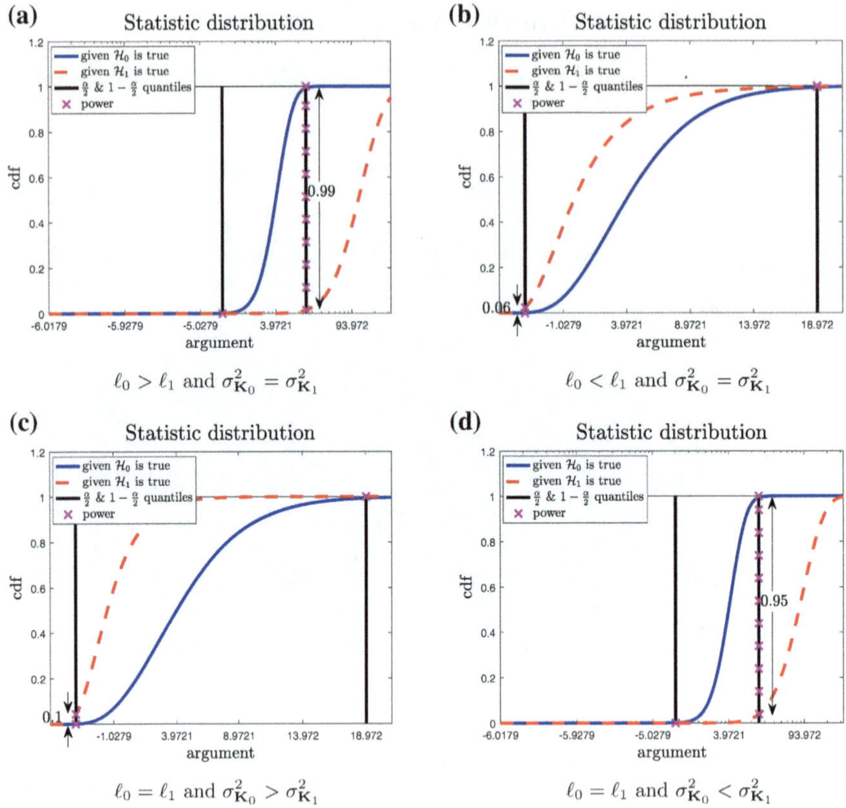

Fig. 5.3 Cdfs of test statistic for the GP-OLCDT with the known hyperparameter vector ϑ under the $\mathcal{H}_0$ and $\mathcal{H}_1$ hypotheses. The vertical lines correspond to $\frac{\alpha_{\text{stat}}}{2}$ and $1 - \frac{\alpha_{\text{stat}}}{2}$ quantiles of the the test statistic distribution under the null hypothesis $\mathcal{H}_0$, the significance level α_{stat} is set to 0.01. These quantiles are used as lower and upper bounds for the test statistic to make a decision of rejecting $\mathcal{H}_0$. The power of the test values are also marked in the plots

For all the GP-OLCDT tests the significance level is set to $\alpha_{\text{stat}} = 0.01$ and sliding window width is set to $L = 10$.

The proposed tests are also compared with the stationary GP Bayesian Online Change Point Detection (GP-BOCPD) algorithm [5]. In the stationary GP-BOCPD method the data y_τ is also assumed to be generated by a GP time series model. In contrast to the proposed framework the functions f_0 and f_1 before and after a change are assumed to be different realisations of the same GP prior whereas in the proposed generative model the hyperparameters of the GP are different before and after the change.

The mean function in GP-BOCPD is also set to zero. The covariance function is the sum of a squared exponential, a periodic, a constant, and "white noise" covariance functions [1]. Within the GP-BOCPD framework change points are estimated via a

Table 5.1 Scenarios characteristics for data simulated by the proposed generative model

Scenario	Length scale, ℓ_1	Signal variance, $\sigma^2_{\mathbf{K}_1}$	Power of the GP-OLCDT with known hyperparameters
1	1	1	0.98996
2	20	1	0.056946
3	3	0.3	0.10001
4	3	4	0.95012

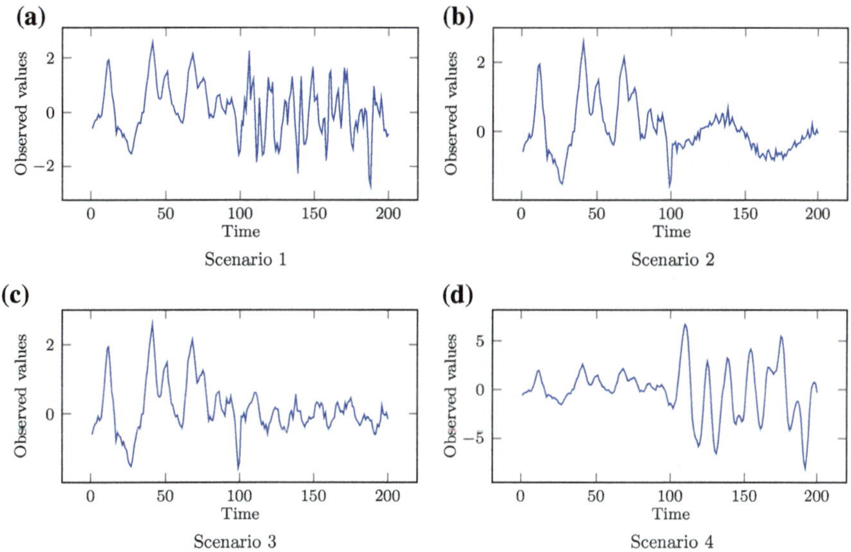

Fig. 5.4 Sample observations generated under 4 scenarios by the proposed generative model

posterior probability of a run-length, i.e., the length of a time period since the last change point. An alarm is generated if the posterior probability of a change point at a considered time moment is more than a threshold. Here the threshold is set to 0.99.

The value of the hyperparameter vector $\boldsymbol{\vartheta}_0$ is fixed for all four scenarios: $\ell_0 = 3$ and $\sigma^2_{\mathbf{K}_0} = 1$. The value for the hyperparameter vector $\boldsymbol{\vartheta}_1$ after the change are provided in Table 5.1. Table 5.1 also presents a power of the GP-OLCDT with known hyperparameters for each scenario. The example of data generated for each of the scenarios can be found in Fig. 5.4.

Consider the average performance of all the methods based on 100 Monte Carlo simulations for each of the scenarios. Table 5.2 presents the obtained values for the performance measures. Here the first alarm generated after the true change time is treated as a correct detection, whereas other alarms are processed as false alarms.

The empirical performance of the GP-OLCDT in terms of the missed detection rate is superior than the theoretical power estimates for the 2-nd and 3-rd scenarios (the missed detection rate can be considered as a complement to the power of a test).

Table 5.2 Change point detection performance on data simulated by the proposed generative model. MTFA—mean time between false alarms, MDFD—mean delay for detection, MDR—missed detection rate

Scenario	Method	MTFA	MDFD	MDR	Power
1	GP-OLCDT with known ϑ_0	157.67	1.31	**0.00**	0.99
	GP-OLCDT with batch estimated ϑ_0	164.53	1.24	**0.00**	0.99 ± 0.00
	GP-OLCDT with sliding estimated ϑ_0	48.19	1.56	0.01	0.99 ± 0.01
	GP-lLRT	**195.95**	**0.60**	0.04	–
	GP-glLRT	177.75	0.65	0.19	–
	GP-BOCPD	129.65	2.40	**0.00**	–
2	GP-OLCDT with known ϑ_0	157.67	11.50	0.10	0.06
	GP-OLCDT with batch estimated ϑ_0	164.53	10.83	0.13	0.06 ± 0.02
	GP-OLCDT with sliding estimated ϑ_0	86.18	8.82	0.23	0.11 ± 0.04
	GP-lLRT	190.88	**0.15**	0.09	–
	GP-glLRT	**192.14**	25.80	**0.07**	–
	GP-BOCPD	190.56	11.52	0.37	–
3	GP-OLCDT with known ϑ_0	157.67	13.63	**0.02**	0.10
	GP-OLCDT with batch estimated ϑ_0	164.53	14.16	0.08	0.10 ± 0.04
	GP-OLCDT with sliding estimated ϑ_0	108.36	17.11	0.30	0.09 ± 0.04
	GP-lLRT	**200.00**	**0.00**	0.98	–
	GP-glLRT	197.43	**0.00**	0.86	–
	GP-BOCPD	195.65	13.28	0.50	–
4	GP-OLCDT with known ϑ_0	157.67	0.64	**0.00**	0.95
	GP-OLCDT with batch estimated ϑ_0	164.53	0.56	**0.00**	0.96 ± 0.02
	GP-OLCDT with sliding estimated ϑ_0	68.23	0.61	0.03	0.96 ± 0.02
	GP-lLRT	**196.91**	0.30	0.03	–
	GP-glLRT	189.99	**0.13**	0.09	–
	GP-BOCPD	188.13	2.14	**0.00**	–

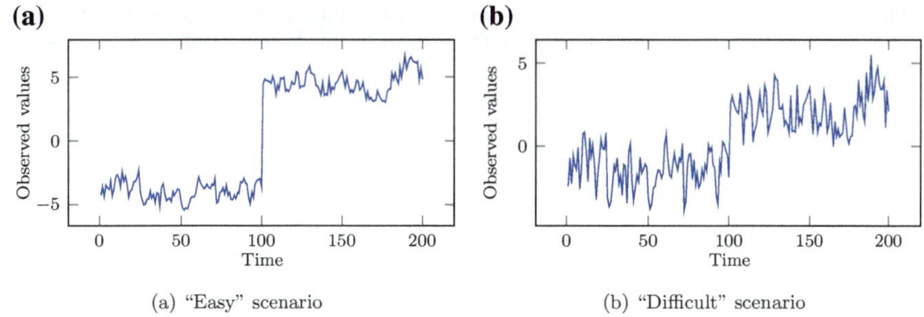

Fig. 5.5 Sample observations generated under 2 scenarios by the GP-BOCPD generative model

In the 1-st and 4-th scenarios the empirical missed detection rate results of the GP-OLCDT confirm the theoretical power performance. The analysis of the results given in Table 5.2 shows though that the 2-nd and 3-rd scenarios are much more difficult for the analysis for all the methods in comparison to the other two scenarios.

The GP-OLCDTs demonstrate better results than the GP-BOCPD in terms of the delay for detection and comparable outputs with respect to both the time between false alarms and the missed detection rate for the 1-st and 4-th scenarios. In the 2-nd and 3-rd scenarios the GP-OLCDTs outperform the GP-BOCPD in terms of the missed detection rate having the similar results for both the time between false alarms and the delay for detection. Although the GP-OLCDT with sliding estimated ϑ_0 has a slightly larger number of false alarms causing lower values of the corresponding performance measure than the other methods. Likelihood ratio tests enable to find the most accurate estimates of change times in all the cases.

5.4.2 *Data Simulated by the GP-BOCPD Model*

To make a fair comparison the methods are also applied on data generated by the GP-BOCPD model, i.e., $\mathbf{f}_{1:100}$ and $\mathbf{f}_{101:200}$ are different realisations of the same GP with fixed hyperparameters.

For data generation the same settings for the mean and covariance functions are used as in the GP-BOCPD algorithm in Sect. 5.4.1.

Although the proposed methods are not limited to any particular type of covariance functions, the squared exponential covariance function is used in the tests. As the data is generated based on the other model rather than the one used in the test assumptions the GP-OLCDT with known hyperparameters and GP-lLRT are not applicable in this setting.

Two scenarios are considered: (1) an "easy" one, where the changes are well distinguishable, and (2) a "difficult" one, where the changes are not easily detected. Figure 5.5 presents sample data obtained under both scenarios.

Table 5.3 Change point detection performance on data simulated by the GP-BOCPD model. MTFA—mean time between false alarms, MDFD—mean delay for detection, MDR—missed detection rate

Scenario	Method	MTFA	MDFD	MDR
"Easy"	GP-OLCDT with batch estimated ϑ_0	151.37	6.59	**0.05**
	GP-OLCDT with sliding estimated ϑ_0	94.48	4.25	0.24
	GP-glLRT	181.19	**0.87**	0.30
	GP-BOCPD	**197.85**	10.13	0.25
"Difficult"	GP-OLCDT with batch estimated ϑ_0	158.93	15.34	**0.17**
	GP-OLCDT with sliding estimated ϑ_0	144.66	20.24	0.54
	GP-glLRT	162.76	**4.08**	0.64
	GP-BOCPD	**191.49**	32.73	0.60

Average performance results among 100 Monte Carlo runs can be found in Table 5.3. Although the proposed methods use the data model assumptions different to the actual generative model, they demonstrate better results than the GP-BOCPD in terms of both the delay for detection and the missed detection rate for the two scenarios. However, the proposed methods generate more false alarms than the GP-BOCPD that leads to the lower performance in terms of the time between false alarms.

5.5 Numerical Experiments with Real Data

In this section the proposed framework is applied to the honey bee dance video data [6]. Honey bees perform a dance as a way to communicate with each other and transfer knowledge about locations of food sources. The dance can be decomposed into three phases: "left turn", "right turn", and "waggle" (Fig. 5.6). The goal is to find change points between the dance phases (stationary behaviours). The original video data provided by the authors [6] is pre-processed, where a dancer bee is tracked and its space coordinates and body angle are extracted. The data consists of 6 sequences.

In this work only spatial coordinates of a tracked bee are processed to find change points between different dance phases. To combine the output from both x and y coordinates of the dancer bee, the methods are applied to both time series independently and then the union of alarms from both time series is used as a resulting output of the method. A function f sampled from a vanilla GP prior is a scalar function, i.e., $f : \mathbb{R}^n \rightarrow \mathbb{R}$. Although different methods [7–9] have been proposed to overcome this constraint, here scalar-value functions are considered.

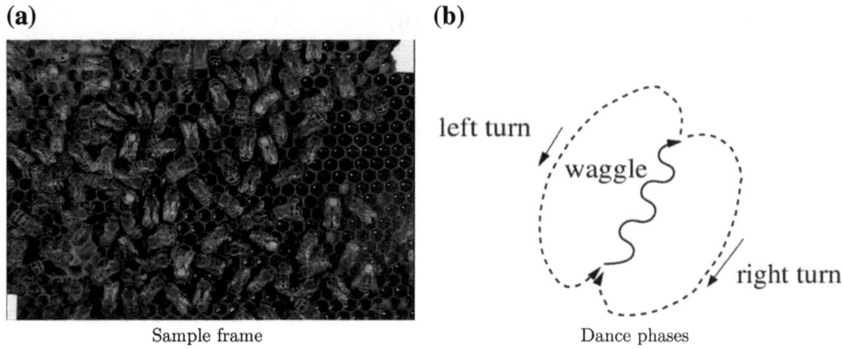

Fig. 5.6 Honey bee dance data. **a** is a sample frame of the original video. **b** is a scheme of dance phases

Since the true data model is not known and multiple change points are expected to be detected, the GP-OLCDT with sliding estimated hyperparameters is used. The following settings are applied: a constant zero mean function, as input time series are centralised and normalised in advance; a squared exponential covariance function. The sliding window width $\tilde{L}$ for hyperparameter estimation is set to 25, and the sliding window width L of data for test statistic computation is set to 5. The significance level $\alpha_{\text{stat}} = 0.01$ is used. The initial values for optimising hyperparameters are set to zero after every change point detection. If there is no change point detected the values of hyperparameters obtained from the previous time step are used as initial values for the current time step.

The GP-BOCPD method is used with the following settings: a constant zero mean function; a covariance function that is a sum of a rational quadratic, constant and "white noise" covariance functions [9]. The first 250 data points are used for hyperparameter learning in each sequence. The threshold value for alarm generation is set to 0.99.

We also compare the method based on penalised contrasts [10, 11][1] (referred below as "Contrast"). This is an offline change point detection algorithm. The number of change points and their locations are determined minimising a so-called contrast function. The method is set to detect changes in the mean and the slope and the maximum number of change points is equal to the true number of change points.

The nearest alarm within 5 time steps before and 15 time steps after a true change point is treated as a correct detection. All other alarms are processed as false alarms. Samples of the input data and change points detected by the methods are presented in Fig. 5.7. The obtained performance measures for all 6 sequences are given in Table 5.4.

Overall results show that the Contrast algorithm generates more alarms leading to low values of all three measures. The GP-BOCPD method raises fewer alarms and therefore has a good performance in terms of the false alarm measure but also has a

[1] The implementation in Matlab 2016a is used (the function findchangepts).

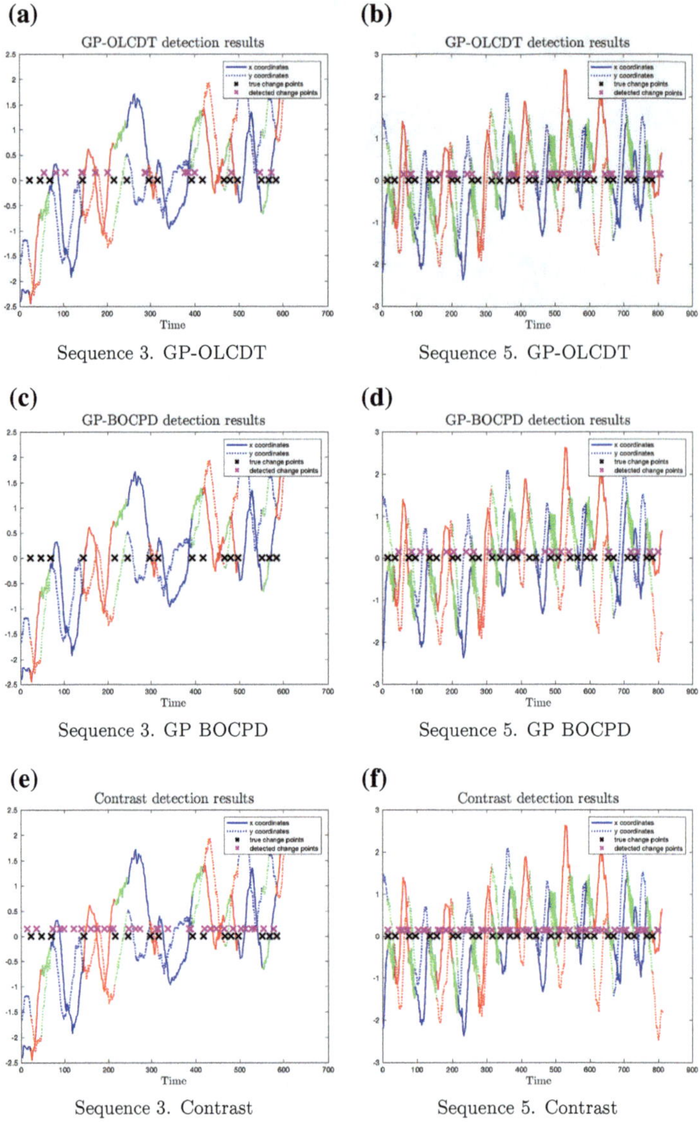

Fig. 5.7 Change point detection on honey bee dance data. Columns correspond to data sequences (two out of total six sequences). The first row presents detections by the proposed GP-OLCDT, the second row presents detections by the GP-BOCPD method, the third row presents detections by the Contrast method. Colours of the data plots represent the true dance phases: blue—"turn left", red—"turn right", green—"waggle"

Table 5.4 Change point detection performance on the bee dance data. MTFA—mean time between false alarms, MDFD—mean delay for detection, MDR—missed detection rate

Sequence	Method	MTFA	MDFD	MDR
1	GP-OLCDT	18.1429	**2**	0.36842
	GP BOCPD	**25.5833**	12	0.84211
	Contrast	18.2174	3.4	**0.21053**
2	GP-OLCDT	19.52	4.1818	0.5
	GP-BOCPD	**31.25**	4.75	0.81818
	Contrast	17.5385	**2.2353**	**0.22727**
3	GP-OLCDT	**28.625**	4.8571	0.5625
	GP-BOCPD	Inf	Inf	1
	Contrast	16.5	**1.9231**	**0.1875**
4	GP-OLCDT	12.3846	3.5385	**0.23529**
	GP-BOCPD	**21**	8.5	0.64706
	Contrast	11.5455	**3**	0.29412
5	GP-OLCDT	9.2609	3	0.32143
	GP-BOCPD	**23.6667**	6.6875	0.42857
	Contrast	10.2333	**1.3846**	**0.071429**
6	GP-OLCDT	**15.9**	4.5	0.6
	GP-BOCPD	Inf	Inf	1
	Contrast	11.7059	**−0.076923**	**0.13333**

very high proportion of missed detections. The proposed GP-OLCDT demonstrates a tradeoff between the false alarm and missed detection rates. In sequences 3 and 6 the GP-BOCPD fails to detect any changes while the GP-OLCDT is able to detect around 40% of change points. In sequences 1 and 4 the GP-OLCDT demonstrates similar results to the offline Contrast method. The analysis of the results given in Fig. 5.7 confirms that the proposed GP-OLCDT generates acceptable number of alarms based on tradeoff between the false alarm and missed detection rates.

5.6 Summary

This chapter presents a general framework for change point detection in time series. A function governing a time series is considered to have a Gaussian process prior. A change in time series is defined as a change in hyperparameters of the Gaussian process, which means a change in a form of functional dependence. The proposed framework can be applied for behaviour analysis and anomaly detection in video, where behaviours are understood as periods with a stationary data distribution between change points, and the change points indicate anomalies as transition from one behaviour to another.

A change point detection problem is formulated within the statistical hypothesis testing framework. We propose likelihood ratio based tests for both cases with known and unknown hyperparameters of a Gaussian process before and after a change. For online change point detection, likelihood-based tests operating within a sliding window are designed.

The theoretical properties of the developed methods are analysed based on the derived probability distribution of the proposed test statistic.

The developed methods are evaluated on both synthetic and real data. Under complicated scenarios of the synthetic data the methods demonstrate significantly better empirical performance than the theoretical estimates. On synthetic data that violates the assumptions applied in the tests the proposed methods outperform the method, which generative model is used in data simulation. The proposed methods are shown to detect changes on real data between different behaviours of a dancing bee where another GP-based change point detection algorithm fails to find any changes. In the sequences where the other GP-based algorithm finds change points the proposed one demonstrates 52% improvement in the mean delay between the true change time and the time of detection.

The next chapter concludes the thesis and indicates directions for future work.

References

1. C.E. Rasmussen, C.K.I. Williams, *Gaussian Processes for Machine Learning* (MIT Press, New York, 2006)
2. F. Gustafsson, *Adaptive filtering and change detection*, vol. 1 (Wiley, New York, 2000)
3. M. Basseville, I.V. Nikiforov, *Detection of Abrupt Changes - Theory and Application* (Prentice Hall, Inc., New York , 1993)
4. R.B. Davies, Algorithm AS 155: The distribution of a linear combination of χ^2 random variables. J. Roy. Stat. Soc. Ser. C (Appl. Stat.) **29**(3), 323–333 (1980)
5. Y. Saatçi, R.D. Turner, C.E. Rasmussen, Gaussian process change point models, in *Proceedings of the 27th International Conference on Machine Learning*, pp. 927–934, June 2010
6. S.M. Oh, J.M. Rehg, T. Balch, F. Dellaert, Learning and inferring motion patterns using parametric segmental switching linear dynamic systems. Int. J. Comput. Vis. (IJCV) **77**(1), 103–124 (2008)
7. M.A. Alvarez, N.D. Lawrence, Computationally efficient convolved multiple output Gaussian processes. J. Mach. Learn. Res. **12**, 1459–1500 (2011)
8. T.V. Nguyen, E. Bonilla, Collaborative multi-output Gaussian processes, in *Proceedings of the 30th Conference on Uncertainty in Artificial Intelligence (UAI 2014)*, pp. 633–643, July 2014
9. R.D. Turner, *Gaussian Processes for State Space Models and Change Point Detection*. Ph.D. thesis, University of Cambridge (2011)
10. M. Lavielle, Using penalized contrasts for the change-point problem. Sig. Process. **85**(8), 1501–1510 (2005)
11. R. Killick, P. Fearnhead, I.A. Eckley, Optimal detection of changepoints with a linear computational cost. J. Am. Stat. Assoc. **107**(500), 1590–1598 (2012)

Chapter 6
Conclusions and Future Work

This chapter provides an overview of the statistical methods and principal contributions presented in this thesis followed by an outline of open research directions for future work.

6.1 Summary of Methods and Contributions

Behaviour analysis and anomaly detection in video are essential parts towards passing the Turing test in computer vision. Machine learning methods show promising results in the context of artificial intelligence in various areas. This thesis develops machine learning methods for unsupervised video processing that extract semantic patterns from data. These patterns are then used for autonomous decision making in anomaly detection. All the methods presented in the thesis can be used for online data processing, which makes them applicable in proactive video analytics systems.

The two approaches for behaviour analysis and anomaly detection are examined in this thesis. In the first part of the thesis extraction of typical local motion patterns and detection of abnormal events, which cannot be explained by these patterns, are considered in the context of topic modeling. Here the research is focused on situations where some periodicity is expected in data and the goal is to detect events that do not fit to this routine, there can be novel events or events appearing in the novel order. Examples of such types of data can be jaywalking or vehicle U-turns in traffic surveillance data.

The second part of the thesis is devoted to detection of sudden changes in behaviours by the change point detection methodology. Methods within this approach do not rely on assumptions about periodicity in data, they are rather aimed to work in unknown situations in contrast to methods presented in the first part. These methods are useful, for example, in surveillance to detect panic in crowd behaviour among people in public places such as stadiums or concert halls.

O. Isupova, *Machine Learning Methods for Behaviour Analysis and Anomaly Detection in Video*, Springer Theses, https://doi.org/10.1007/978-3-319-75508-3_6

Chapter 3 presents an examination of a topic model that considers dynamic dependencies between documents. The aim of this chapter is to compare different learning schemes for the topic model. Two novel learning algorithms are developed: based on MAP estimation using the EM-algorithm and variational Bayes inference. The algorithms are thoroughly compared with the existing Gibbs sampling scheme. Two different methods for likelihood approximation used in final anomaly detection decision making are analysed. A more efficient method based on point estimates shows competitive results with a Monte Carlo approximation method. A novel anomaly localisation procedure is proposed. This procedure is elegantly embedded in the topic modeling framework. The empirical results based on real video data confirm the superiority of the developed learning algorithms in terms of anomaly detection performance and competence of the proposed localisation procedure.

The nonparametric perspective of the topic modeling approach for behaviour analysis and anomaly detection is then studied in Chap. 4. A novel Bayesian nonparametric topic model is proposed in this chapter. It is demonstrated that the current state of the art nonparametric topic model—the HDP—is unable to capture abnormal events. Consideration of the dynamics on topic mixtures in documents proposed in the novel model significantly improves the ability of a topic model to detect anomalies. The intuition of the proposed dynamics is based on the fact of motion continuance in real life. An efficient inference algorithm is derived for the model, which is divided into two phases: a conventional batch data processing on training data and an online algorithm for test data processing. A normality measure based on predictive likelihood of a newly observed document is developed for decision making in anomaly detection.

Chapter 5 considers the problem of anomaly detection and behaviour analysis in the context of change point detection. From this perspective periods between change points can be considered as different types of behaviour. If at any moment a *normal* behaviour is expected then a change point, detected after this moment, would indicate an anomaly. Input data is assumed to be in a form of time series with a Gaussian process prior. A general framework for change point detection is proposed in this chapter, where a change is defined as alteration of the Gaussian process hyperparameters. The statistical hypothesis testing approach is employed and several statistical tests are designed and analysed. The statistical hypothesis testing approach allows to derive theoretical estimates of algorithm performance. One of the important benefits of the developed framework is that it avoids a detailed specification of the time series function before and after a change, particular values of hyperparameters are estimated from the data. The evaluation results on both synthetic and real data show that the proposed methods achieve a reasonable tradeoff between false alarms and missed detection rates.

6.2 Suggestions for Future Work

This section discusses of open research directions that arise based on the work presented in this thesis. It also provides an outline of potential application areas outside video processing where the proposed statistical methods can be employed.

6.2.1 *Inference in Topic Modeling*

As it is shown in Chap. 3 different inference algorithms can achieve different results in terms of anomaly detection performance. They also have different computational efficiency.

6.2.1.1 Different Markov Chain Monte Carlo Samplers

In Chap. 4 a Gibbs sampler based on the Chinese restaurant franchise is developed for the dynamic HDP. A split-merge extension of this Gibbs sampler for the standard HDP is shown to improve a convergence rate [1]. In the original work on HDP [2] two other samplers are also proposed: based on augmented representation and on direct assignments, allowing to factorise the posterior across the documents. These Gibbs samplers can suffer from slow convergence on sequential data since successive data points have a strong mutual dependence [3]. To overcome this issue several methods have been proposed, including a blocked Gibbs sampler [4] and beam sampler [5]. The implementation of these samplers for the proposed dynamic HDP might improve its performance.

6.2.1.2 Variational Inference

Another direction in exploring inference algorithms for the proposed dynamic HDP is variational inference [6]. Variational algorithms can be developed for online data processing [7], which is essential for the anomaly detection problem. Design of an online variational inference algorithm for the proposed dynamic HDP can be based on its counterpart for the standard HDP [8].

6.2.2 *Alternative Dynamics in Topic Modeling*

Chapter 4 demonstrates that the proposed dynamics on topic mixtures make the HDP topic model capable to detect anomalies in video data. The proposed dynamics are based on the evident fact that if an activity is present in the current video clip it is likely to be present in the next clip as well. Although it might be not the only dynamics, which can be incorporated into the HDP topic model to further improve anomaly detection performance.

For example, it can be noticed that local activities, captured in one scene, rarely appear or disappear randomly. The activities rather replace each other. Imagine there is an activity that represents a pedestrian motion on a crosswalk. When pedestrians finish crossing a road, this activity will disappear giving place to another activity that represents a pedestrian motion on a sidewalk. A motion starting in one place of the scene does not usually totally disappear (unless objects leave the scene), it rather continues in another place.

This intuition can be modelled with Markovian dependencies imposed on topic transitions. For example, in [9] an infinite number of infinite hidden Markov models are used for similar ideas. In the hidden Markov models topics are hidden states and visual words are observed variables. These types of models might be expected to better explain complex motion patterns and therefore improve both descriptive behaviour analysis and anomaly detection performance.

6.2.3 Gaussian Process Change Point Detection

Gaussian process change point detection introduced in Chap. 5 represents a promising approach in detection theory. It leads to a lot of avenues for further research, which are outlined below.

The methods presented in Chap. 5 are directly applicable to autoregressive models. It is worth exploring the performance of the methods with autoregressive data model on real data. For the related Gaussian process change point detection method that employed Bayesian inference it is shown that the autoregressive data model often better explains real data than the Gaussian process time series model [10].

As demonstrated in Sect. 5.4.1 there are data settings, in which the proposed statistical test is theoretically proven to have a low performance. These specific data settings represent situations when a data model before a change is more flexible than a data model after the change, therefore data after the change is well described by the first data model. One of the possible solutions can be imposing a hyperprior on the Gaussian process hyperparameters. This prior should help to distinguish the distribution of the test statistic under models before and after a change.

Straightforward inference in Gaussian processes has a cubic (with respect to the number of training points) computational complexity due to inversion of a covariance matrix. In the proposed methods the number of training points is bounded by the width of the sliding window used in the test statistic computation. To ensure quickest change point detection the width of this sliding window is kept small. Therefore, the cubic complexity is not an issue as the number of training points is small. However, the width of the sliding window using for hyperparameter estimation can be large to obtain accurate estimates of the hyperparameters and methods for efficient Gaussian process inference can be studied for it. For example, Kalman filtering reformulation of the Gaussian process regression problem [11] has a linear computational complexity and it then represents a promising approach.

In Chap. 5 only a standard single output Gaussian process is considered, that leads that only one-dimensional time series data is analysed. Multiple output Gaussian processes [12–14] can be developed within the proposed framework for change point detection to process multidimensional data.

Gaussian processes are not the only way to represent the prior knowledge. A Student-t process is proposed as a promising alternative to Gaussian processes [15]. Student-t process data can be studied in terms of applicability of the ideas proposed in Chap. 5 for change point detection.

6.2.4 *Potential Applications of the Proposed Statistical Methods*

While throughout the thesis the developed statistical methods are considered in the context of behaviour analysis and anomaly detection in video, they can be used in a wide range of applications.

The Markov clustering topic model can be applied to any kind of data, where repeated behaviours are expected. It means that any data affected by seasonal variations can be processed with the Markov clustering topic model, e.g., retail transactions or traffic logs in computer systems. The model can be used for extracting typical patterns from data. It can also be applied for anomaly detection, e.g., a fraud in retail or an intrusion attack on a computer system. The proposed learning methods are then expected to improve descriptive analysis of the typical patterns and anomaly detection performance.

The proposed dynamic HDP topic model can be used in text mining to analyse time-stamped documents, such as news streams, tweets or scientific papers. This model can be used for better data understanding or for anomaly detection, such as detection of atypical trends in a social network or a novel area detection in scientific papers, for example, a document with unusual topic mixture can indicate a pioneering work.

The proposed change point detection framework is a general tool and can be applied to any time series data. An outline of the possible applications is presented in Sect. 2.4.1.

References

1. C. Wang, D.M. Blei, A split-merge MCMC algorithm for the hierarchical Dirichlet process. arXiv:1201.1657 (2012)
2. Y.W. Teh, M.I. Jordan, M.J. Beal, D.M. Blei, Hierarchical Dirichlet processes. J. Am. Stat. Assoc. **101**(476), 1566–1581 (2006)

3. S.L. Scott, Bayesian methods for hidden Markov models: recursive computing in the 21st century. J. Am. Stat. Assoc. **97**(457), 337–351 (2002)
4. E.B. Fox, E.B. Sudderth, M.I. Jordan, A.S. Willsky, A sticky HDP-HMM with application to speaker diarization. Ann. Appl. Stat. **5**(2A), 1020–1056 (2011)
5. J. Van Gael, Y. Saatci, Y.W. Teh, Z. Ghahramani, Beam sampling for the infinite hidden Markov model, in *Proceedings of the 25th International Conference on Machine Learning* (2008), pp. 1088–1095
6. M. Jordan, Z. Ghahramani, T. Jaakkola, L. Saul, An introduction to variational methods for graphical models. Mach. Learn. **37**(2), 183–233 (1999)
7. M.D. Hoffman, D.M. Blei, C. Wang, J.W. Paisley, Stochastic variational inference. J. Mach. Learn. Res. **14**(1), 1303–1347 (2013)
8. C. Wang, J.W. Paisley, D.M. Blei, Online variational inference for the hierarchical Dirichlet process, in *Proceedings of the 14th International Conference on Artificial Intelligence and Statistics (AISTATS 2011)* (2011), pp. 752–760
9. D. Kuettel, M. Breitenstein, L. Van Gool, V. Ferrari, What's going on? Discovering spatio-temporal dependencies in dynamic scenes, in *Proceedings of the 2010 IEEE Conference on Computer Vision and Pattern Recognition (CVPR)* (2010), pp. 1951–1958
10. Y. Saatçi, R.D. Turner, C.E. Rasmussen, Gaussian process change point models, in *Proceedings of the 27th International Conference on Machine Learning* (2010), pp. 927–934
11. S. Särkkä, J. Hartikainen, Infinite-dimensional Kalman filtering approach to spatio-temporal Gaussian process regression, in *Proceedings of the 15th International Conference on Artificial Intelligence and Statistics (AISTATS 2012)* (2012), pp. 993–1001
12. M.A. Alvarez, N.D. Lawrence, Computationally efficient convolved multiple output Gaussian processes. J. Mach. Learn. Res. **12**, 1459–1500 (2011)
13. T.V. Nguyen, E. Bonilla, Collaborative multi-output Gaussian processes, in *Proceedings of the 30th Conference on Uncertainty in Artificial Intelligence (UAI 2014)* (2014), pp. 633–643
14. R.D. Turner, Gaussian processes for state space models and change point detection. Ph.D. thesis, University of Cambridge, 2011
15. A. Shah, A.G. Wilson, Z. Ghahramani, Student-t processes as alternatives to Gaussian processes, in *Proceedings of the 17th International Conference on Artificial Intelligence and Statistics (AISTATS 2014)* (2014), pp. 877–885

Appendix A
EM for MCTM Derivation

This appendix provides the details of the derivation of the proposed EM learning algorithm for the MCTM presented in Chap. 3.

The objective function in the EM-algorithm is:

$$\mathcal{Q}(\boldsymbol{\Omega}, \boldsymbol{\Omega}^{\text{old}}) + \log p(\boldsymbol{\Omega}|\boldsymbol{\eta}, \boldsymbol{\alpha}, \boldsymbol{\varkappa}, \boldsymbol{\upsilon}) \overset{\text{eq. (3.4)}}{=} \mathbb{E}_{p(\mathbf{z}_{1:J_{tr}}, \mathbf{b}_{1:J_{tr}}|\mathbf{w}_{1:J_{tr}}, \boldsymbol{\Omega}^{\text{old}})} \log p(\mathbf{w}_{1:J_{tr}}, \mathbf{z}_{1:J_{tr}}, \mathbf{b}_{1:J_{tr}}|\boldsymbol{\Omega}) + \log p(\boldsymbol{\Omega}|\boldsymbol{\eta}, \boldsymbol{\alpha}, \boldsymbol{\varkappa}, \boldsymbol{\upsilon}) \tag{A.1}$$

Since both $\mathbf{z}_{1:J_{tr}}$ and $\mathbf{b}_{1:J_{tr}}$ are discrete, the expected value is a sum over probable values and (A.1) is given as:

$$\mathbb{E}_{p(\mathbf{z}_{1:J_{tr}}, \mathbf{b}_{1:J_{tr}}|\mathbf{w}_{1:J_{tr}}, \boldsymbol{\Omega}^{\text{old}})} \log p(\mathbf{w}_{1:J_{tr}}, \mathbf{z}_{1:J_{tr}}, \mathbf{b}_{1:J_{tr}}|\boldsymbol{\Omega}) + \log p(\boldsymbol{\Omega}|\boldsymbol{\eta}, \boldsymbol{\alpha}, \boldsymbol{\varkappa}, \boldsymbol{\upsilon}) = \sum_{\mathbf{z}_{1:J_{tr}}} \sum_{\mathbf{b}_{1:J_{tr}}} \left(p(\mathbf{z}_{1:J_{tr}}, \mathbf{b}_{1:J_{tr}}|\mathbf{w}_{1:J_{tr}}, \boldsymbol{\Omega}^{Old}) \log p(\mathbf{w}_{1:J_{tr}}, \mathbf{z}_{1:J_{tr}}, \mathbf{b}_{1:J_{tr}}|\boldsymbol{\Omega}, \boldsymbol{\alpha}, \boldsymbol{\eta}, \boldsymbol{\upsilon}, \boldsymbol{\varkappa})\right) + \log p(\boldsymbol{\Omega}|\boldsymbol{\eta}, \boldsymbol{\alpha}, \boldsymbol{\varkappa}, \boldsymbol{\upsilon}) \tag{A.2}$$

Substituting the expression for the full likelihood from (3.2) into (A.2) and (A.2) into (A.1) and marginalising the hidden variables, we can write the objective function in the EM-algorithm as:

$$\mathcal{Q}(\boldsymbol{\Omega}, \boldsymbol{\Omega}^{\text{old}}) + \log p(\boldsymbol{\Omega}|\boldsymbol{\eta}, \boldsymbol{\alpha}, \boldsymbol{\varkappa}, \boldsymbol{\upsilon}) = Const + \sum_{b_1 \in \mathcal{B}} \left(\log \omega_{b_1}\, p(b_1|\mathbf{w}_{1:J_{tr}}, \boldsymbol{\Omega}^{Old})\right) + \sum_{j=2}^{J_{tr}} \sum_{b_j \in \mathcal{B}} \sum_{b_{j-1} \in \mathcal{B}} \left(\log \xi_{b_j b_{j-1}}\, p(b_j, b_{j-1}|\mathbf{w}_{1:J_{tr}}, \boldsymbol{\Omega}^{Old})\right) +$$

O. Isupova, *Machine Learning Methods for Behaviour Analysis and Anomaly Detection in Video*, Springer Theses, https://doi.org/10.1007/978-3-319-75508-3

$$\sum_{j=1}^{J_{tr}} \sum_{i=1}^{N_j} \sum_{z_{ji} \in \mathcal{K}} \left(\log \phi_{w_{ji}\, z_{ji}}\; p(z_{ji} | \mathbf{w}_{1:J_{tr}}, \boldsymbol{\Omega}^{Old})\right) +$$
$$\sum_{j=1}^{J_{tr}} \sum_{i=1}^{N_j} \sum_{b_j \in \mathcal{B}} \sum_{z_{ji} \in \mathcal{K}} \left(\log \theta_{z_{ji}\, b_j}\; p(z_{ji}, b_j | \mathbf{w}_{1:J_{tr}}, \boldsymbol{\Omega}^{Old})\right) +$$
$$\sum_{b \in \mathcal{B}} (\varkappa_b - 1) \log \omega_b + \sum_{b \in \mathcal{B}} \sum_{b' \in \mathcal{B}} (\upsilon_b - 1) \log \xi_{bb'} +$$
$$\sum_{b \in \mathcal{B}} \sum_{k \in \mathcal{K}} (\alpha_k - 1) \log \theta_{kb} + \sum_{k \in \mathcal{K}} \sum_{w \in \mathcal{V}} (\eta_w - 1) \log \phi_{wk} \tag{A.3}$$

During the M-step the function (A.3) is maximised with respect to the parameters $\boldsymbol{\Omega}$ with fixed values for $p(b_1 | \mathbf{w}_{1:J_{tr}}, \boldsymbol{\Omega}^{Old})$, $p(b_j, b_{j-1} | \mathbf{w}_{1:J_{tr}}, \boldsymbol{\Omega}^{Old})$, $p(z_{ji} | \mathbf{w}_{1:J_{tr}}, \boldsymbol{\Omega}^{Old})$, $p(z_{ji}, b_j | \mathbf{w}_{1:J_{tr}}, \boldsymbol{\Omega}^{Old})$. The optimisation problem can be solved separately for each parameter. Optimisation of all parameters is performed similarly that leads to the Eqs. (3.6)–(3.8), here the details for the update of the parameters $\boldsymbol{\Phi}$ are provided.

There is an optimisation problem for the parameters $\boldsymbol{\Phi}$:

$$\sum_{j=1}^{J_{tr}} \sum_{i=1}^{N_j} \sum_{z_{ji} \in \mathcal{K}} \left(\log \phi_{w_{ji}\, z_{ji}}\; p(z_{ji} | \mathbf{w}_{1:J_{tr}}, \boldsymbol{\Omega}^{Old})\right) + \sum_{k \in \mathcal{K}} \sum_{w \in \mathcal{V}} (\eta_w - 1) \log \phi_{wk} \longrightarrow \max_{\boldsymbol{\Phi}} \tag{A.4}$$

and constraints:

$$\sum_{w \in \mathcal{V}} \phi_{wk} = 1, \quad \forall k \in \mathcal{K} \tag{A.5}$$

that ensure the columns of the matrix $\boldsymbol{\Phi}$ form valid probability distribution vectors.

A Lagrangian for the problem (A.4)–(A.5) is given as:

$$\mathcal{L} = \sum_{j=1}^{J_{tr}} \sum_{i=1}^{N_j} \sum_{z_{ji} \in \mathcal{K}} \left(\log \phi_{w_{ji}\, z_{ji}}\; p(z_{ji} | \mathbf{w}_{1:J_{tr}}, \boldsymbol{\Omega}^{Old})\right) +$$
$$\sum_{k \in \mathcal{K}} \sum_{w \in \mathcal{V}} (\eta_w - 1) \log \phi_{wk} - \sum_{k \in \mathcal{K}} \lambda_{\phi_k} \left(\sum_{w \in \mathcal{V}} \phi_{wk} - 1 \right) \longrightarrow \max_{\boldsymbol{\Phi}}, \tag{A.6}$$

where λ_{ϕ_k} are Lagrange multipliers.

To find Lagrange multipliers and parameter values ϕ_{wk} we equate the derivative of the Lagrangian (A.6) to zero:

$$\frac{\partial \mathcal{L}}{\partial \phi_{wk}} = \frac{\overbrace{\sum_{j=1}^{J_{tr}} \sum_{i=1}^{N_j} p(z_{ji} = k | \mathbf{w}_{1:J_{tr}}, \mathbf{\Omega}^{Old}) \mathbb{1}(w_{ji} = w)}^{\hat{l}_{wk}^{\text{EM}}} + \eta_w - 1}{\phi_{wk}} - \lambda_{\phi_k} = 0 \quad \text{(A.7)}$$

$$\lambda_{\phi_k} \phi_{wk} = \eta_w + \hat{l}_{wk}^{\text{EM}} - 1 \quad \text{(A.8)}$$

We sum over w both sides and write (A.8) as:

$$\lambda_{\phi_k} = \sum_{w' \in \mathcal{V}} \left(\eta_{w'} + \hat{l}_{w'k}^{\text{EM}} - 1 \right) \quad \text{(A.9)}$$

Substituting the expression for λ_{ϕ_k} (A.9) into (A.8) we get the formula (3.6), where the operation $(\cdot)_+$ is applied to ensure non-negativity of obtained values for ϕ_{wk}.

During the E-step $p(b_1 | \mathbf{w}_{1:J_{tr}}, \mathbf{\Omega}^{Old})$, $p(b_j, b_{j-1} | \mathbf{w}_{1:J_{tr}}, \mathbf{\Omega}^{Old})$, $p(z_{ji} | \mathbf{w}_{1:J_{tr}}, \mathbf{\Omega}^{Old})$, $p(z_{ji}, b_j | \mathbf{w}_{1:J_{tr}}, \mathbf{\Omega}^{Old})$ are updated with fixed values for the parameters. For the efficient implementation the forward-backward steps are developed for the auxiliary variables $\acute{\alpha}_b(j)$ and $\acute{\beta}_b(j)$:

$$\acute{\alpha}_b(j) \stackrel{\text{def}}{=} p(\mathbf{w}_1, \ldots, \mathbf{w}_j, b_j = b | \mathbf{\Omega}^{Old}) =$$

$$\sum_{\mathbf{b}_{1:j-1}} \omega_{b_1}^{Old} \left[\prod_{j'=2}^{j-1} \xi_{b_{j'} b_{j'-1}}^{Old} \right] \left[\prod_{j'=1}^{j-1} \prod_{i=1}^{N_{j'}} \sum_{k \in \mathcal{K}} \phi_{w_{j'i} k}^{Old} \theta_{k b_{j'}}^{Old} \right] \xi_{b_j = b\, b_{j-1}}^{Old} \prod_{i=1}^{N_j} \sum_{k \in \mathcal{K}} \phi_{w_{ji} k}^{Old} \theta_{k\, b_j = b}^{Old} \quad \text{(A.10)}$$

Reorganisation of the terms in (A.10) leads to the recursive expressions (3.10).

Similarly for $\acute{\beta}_b(j)$:

$$\acute{\beta}_b(j) \stackrel{\text{def}}{=} p(\mathbf{w}_{j+1}, \ldots, \mathbf{w}_{J_{tr}} | b_j = b, \mathbf{\Omega}^{Old}) =$$

$$\sum_{\mathbf{b}_{j+1:J_{tr}}} \xi_{b_{j+1} b_j = b}^{Old} \left[\prod_{j'=j+2}^{J_{tr}} \xi_{b_{j'} b_{j'-1}}^{Old} \right] \prod_{j'=j+1}^{J_{tr}} \prod_{i=1}^{N_{j'}} \sum_{k \in \mathcal{K}} \phi_{w_{j'i} k}^{Old} \theta_{k b_{j'}}^{Old}. \quad \text{(A.11)}$$

The recursive formula (3.11) is obtained by interchanging the terms in (A.11).

The required posterior of the hidden variable terms $p(b_1|\mathbf{w}_{1:J_{tr}}, \mathbf{\Omega}^{Old})$, $p(b_j, b_{j-1}|\mathbf{w}_{1:J_{tr}}, \mathbf{\Omega}^{Old})$, $p(z_{ji}|\mathbf{w}_{1:J_{tr}}, \mathbf{\Omega}^{Old})$, $p(z_{ji}, b_j|\mathbf{w}_{1:J_{tr}}, \mathbf{\Omega}^{Old})$ are then expressed via the auxiliary variables $\acute{\alpha}_b(j)$ and $\acute{\beta}_b(j)$, which leads to (3.13)–(3.16). The details of derivation of (3.14) are provided here. The other formulae can be obtained similarly.

According to the definition of conditional probability $p(b_j, b_{j-1}|\mathbf{w}_{1:J_{tr}}, \mathbf{\Omega}^{Old})$ can be expressed as:

$$p(b_j, b_{j-1}|\mathbf{w}_{1:J_{tr}}, \mathbf{\Omega}^{Old}) = \frac{p(b_j, b_{j-1}, \mathbf{w}_{1:J_{tr}}, \mathbf{\Omega}^{Old})}{\underbrace{p(\mathbf{w}_{1:J_{tr}}, \mathbf{\Omega}^{Old})}_{Z}} \tag{A.12}$$

The product rule allows to further rewrite the numerator in (A.12):

$$\frac{1}{Z}p(b_j, b_{j-1}, \mathbf{w}_{1:J_{tr}}, \mathbf{\Omega}^{Old}) = \frac{1}{Z}p(\mathbf{w}_{j+1:J_{tr}}|\cancel{\mathbf{w}_{1:j}}, b_j, \cancel{b_{j-1}}, \mathbf{\Omega}^{Old})\times$$
$$p(\mathbf{w}_j|\cancel{\mathbf{w}_{1:j-1}}, b_j, \cancel{b_{j-1}}, \mathbf{\Omega}^{Old})p(b_j|b_{j-1}, \cancel{\mathbf{w}_{1:j-1}}, \mathbf{\Omega}^{Old})p(\mathbf{w}_{1:j-1}, b_{j-1}|\mathbf{\Omega}^{Old}), \tag{A.13}$$

where the terms that are conditionally independent are cancelled out.

The sum rule for the term $p(\mathbf{w}_j|b_j, \mathbf{\Omega}^{Old})$ can be applied and (A.13) can be given as:

$$\frac{1}{Z}p(b_j, b_{j-1}, \mathbf{w}_{1:J_{tr}}, \mathbf{\Omega}^{Old}) = \frac{1}{Z}\underbrace{p(\mathbf{w}_{j+1:J_{tr}}|b_j, \mathbf{\Omega}^{Old})}_{\acute{\beta}_b(j)}\underbrace{p(b_j|b_{j-1}, \mathbf{\Omega}^{Old})}_{\xi^{Old}_{b_j b_{j-1}}}\times$$
$$\underbrace{p(\mathbf{w}_{1:j-1}, b_{j-1}|\mathbf{\Omega}^{Old})}_{\acute{\alpha}_{b_{j-1}}(j-1)}\prod_{i=1}^{N_j}\sum_{k\in\mathcal{K}}\underbrace{p(w_{ji}|z_{ji}=k, \cancel{b_j}, \mathbf{\Omega}^{Old})}_{\phi^{Old}_{w_{ji}k}}\underbrace{p(z_{ji}=k|b_j, \mathbf{\Omega}^{Old})}_{\theta^{Old}_{kb_j}}, \tag{A.14}$$

which leads to the formula (3.14).

Appendix B
VB for MCTM Derivation

The details of the derivation of the proposed variational Bayes learning algorithm for the MCTM presented in Chap. 3 are given in this appendix.

We have separated the parameters and the hidden variables. Let us consider the update formula of the variational Bayes inference scheme [1] for the parameters:

$$\log q(\boldsymbol{\Omega}) = Const + \mathbb{E}_{q(\mathbf{z}_{1:J_{tr}}, \mathbf{b}_{1:J_{tr}})} \log p(\mathbf{w}_{1:J_{tr}}, \mathbf{z}_{1:J_{tr}}, \mathbf{b}_{1:J_{tr}}, \boldsymbol{\Omega} | \boldsymbol{\varkappa}, \boldsymbol{\upsilon}, \boldsymbol{\alpha}, \boldsymbol{\eta}) =$$

$$Const + \mathbb{E}_{q(\mathbf{z}_{1:J_{tr}}, \mathbf{b}_{1:J_{tr}})} \Bigg(\sum_{b \in \mathcal{B}} (\varkappa_b - 1) \log \omega_b + \sum_{b \in \mathcal{B}} \sum_{b' \in \mathcal{B}} \left(\upsilon_{b'} - 1\right) \log \xi_{b' b} +$$

$$\sum_{b \in \mathcal{B}} \sum_{k \in \mathcal{K}} (\alpha_k - 1) \log \theta_{kb} + \sum_{k \in \mathcal{K}} \sum_{w \in \mathcal{V}} (\eta_w - 1) \log \phi_{wk} + \sum_{b \in \mathcal{B}} \mathbb{1}(b_1 = b) \log \omega_b +$$

$$\sum_{j=2}^{J_{tr}} \sum_{b \in \mathcal{B}} \sum_{b' \in \mathcal{B}} \mathbb{1}(b_j = b') \mathbb{1}(b_{j-1} = b) \log \xi_{b' b} + \sum_{j=1}^{J_{tr}} \sum_{i=1}^{N_j} \sum_{k \in \mathcal{K}} \mathbb{1}\left(z_{ji} = k\right) \log \phi_{w_{ji} k} +$$

$$\sum_{j=1}^{J_{tr}} \sum_{i=1}^{N_j} \sum_{b \in \mathcal{B}} \sum_{k \in \mathcal{K}} \mathbb{1}(z_{ji} = k) \mathbb{1}(b_j = b) \log \theta_{kb} \Bigg) \tag{B.1}$$

One can notice that $\log q(\boldsymbol{\Omega})$ is further factorised as in (3.18). Now each factorisation term can be considered independently. Derivations of the Eqs. (3.19)–(3.22) are very similar to each other. We provide the derivation only of the term $q(\boldsymbol{\Phi})$:

$$\log q(\boldsymbol{\Phi}) =$$

$$Const + \mathbb{E}_{q(\mathbf{z}_{1:J_{tr}}, \mathbf{b}_{1:J_{tr}})} \left(\sum_{k \in \mathcal{K}} \sum_{w \in \mathcal{V}} (\eta_w - 1) \log \phi_{wk} + \sum_{j=1}^{J_{tr}} \sum_{i=1}^{N_j} \sum_{k \in \mathcal{K}} \mathbb{1}\left(z_{ji} = k\right) \log \phi_{w_{ji} k} \right) =$$

O. Isupova, *Machine Learning Methods for Behaviour Analysis and Anomaly Detection in Video*, Springer Theses, https://doi.org/10.1007/978-3-319-75508-3

$$Const + \sum_{k\in\mathcal{K}}\sum_{w\in\mathcal{V}} (\eta_w - 1)\log\phi_{wk} + \sum_{j=1}^{J_{tr}}\sum_{i=1}^{N_j}\sum_{k\in\mathcal{K}} \log\phi_{w_{ji}k}\,\underbrace{\mathbb{E}_{q(\mathbf{z}_{1:J_{tr}},\mathbf{b}_{1:J_{tr}})}\left(\mathbb{1}\left(z_{ji}=k\right)\right)}_{q(z_{ji}=k)} =$$

$$Const + \sum_{k\in\mathcal{K}}\sum_{w\in\mathcal{V}} \log\phi_{wk}\left(\eta_w - 1 + \sum_{j=1}^{J_{tr}}\sum_{i=1}^{N_j}\mathbb{1}(w_{ji}=w)q(z_{ji}=k)\right) \quad \text{(B.2)}$$

It can be noticed from (B.2) that the distribution of $\mathbf{\Phi}$ is a product of the Dirichlet distributions (3.19).

The update formula in the variational Bayes inference scheme for the hidden variables is as follows:

$$\log q(\mathbf{z}_{1:J_{tr}},\mathbf{b}_{1:J_{tr}}) = Const + \mathbb{E}_{q(\boldsymbol{\omega})q(\boldsymbol{\Xi})q(\boldsymbol{\Theta})q(\boldsymbol{\Phi})}\log p(\mathbf{w}_{1:J_{tr}},\mathbf{z}_{1:J_{tr}},\mathbf{b}_{1:J_{tr}},\boldsymbol{\Omega}|\boldsymbol{\varkappa},\boldsymbol{\upsilon},\boldsymbol{\alpha},\boldsymbol{\eta}) =$$

$$Const + \sum_{b\in\mathcal{B}}\mathbb{1}\left(b_1=b\right)\mathbb{E}_{q(\boldsymbol{\omega})}\log\omega_b + \sum_{j=2}^{J_{tr}}\sum_{b\in\mathcal{B}}\sum_{b'\in\mathcal{B}}\mathbb{1}\left(b_j=b'\right)\mathbb{1}\left(b_{j-1}=b\right)\mathbb{E}_{q(\boldsymbol{\Xi})}\log\xi_{b'b} +$$

$$\sum_{j=1}^{J_{tr}}\sum_{i=1}^{N_j}\sum_{k\in\mathcal{K}}\mathbb{1}\left(z_{ji}=k\right)\mathbb{E}_{q(\boldsymbol{\Phi})}\log\phi_{w_{ji}k} + \sum_{j=1}^{J_{tr}}\sum_{i=1}^{N_j}\sum_{b\in\mathcal{B}}\sum_{k\in\mathcal{K}}\mathbb{1}\left(z_{ji}=k\right)\mathbb{1}\left(b_j=b\right)\mathbb{E}_{q(\boldsymbol{\Theta})}\log\theta_{kb} \quad \text{(B.3)}$$

We know from the parameter update (3.19)–(3.22) that their distributions are Dirichlet. Therefore, $\mathbb{E}_{q(\boldsymbol{\omega})}\log\omega_b = \psi\left(\tilde{\varkappa}_b\right) - \psi\left(\sum_{b'\in\mathcal{B}}\tilde{\varkappa}_{b'}\right)$ (see, for example, [1]) and similarly for all the other expected value expressions.

Using the introduced notations (3.23)–(3.26) the update formula (B.3) for the hidden variables can be then expressed as:

$$\log q(\mathbf{z}_{1:J_{tr}},\mathbf{b}_{1:J_{tr}}) =$$

$$Const + \sum_{b\in\mathcal{B}}\mathbb{1}\left(b_1=b\right)\log\tilde{\omega}_b + \sum_{j=2}^{J_{tr}}\sum_{b\in\mathcal{B}}\sum_{b'\in\mathcal{B}}\mathbb{1}\left(b_j=b'\right)\mathbb{1}\left(b_{j-1}=b\right)\log\tilde{\xi}_{b'b} +$$

$$\sum_{j=1}^{J_{tr}}\sum_{i=1}^{N_j}\sum_{k\in\mathcal{K}}\mathbb{1}\left(z_{ji}=k\right)\log\tilde{\phi}_{w_{ji}k} + \sum_{j=1}^{J_{tr}}\sum_{i=1}^{N_j}\sum_{b\in\mathcal{B}}\sum_{k\in\mathcal{K}}\mathbb{1}\left(z_{ji}=k\right)\mathbb{1}\left(b_j=b\right)\log\tilde{\theta}_{kb} \quad \text{(B.4)}$$

The approximated distribution of the hidden variables is then:

$$q(\mathbf{z}_{1:J_{tr}},\mathbf{b}_{1:J_{tr}}) = \frac{1}{\tilde{Z}}\tilde{\omega}_{b_1}\left[\prod_{j=2}^{J_{tr}}\tilde{\xi}_{b_j b_{j-1}}\right]\prod_{j=1}^{J_{tr}}\prod_{i=1}^{N_j}\tilde{\phi}_{w_{ji}z_{ji}}\tilde{\theta}_{z_{ji}b_j}, \quad \text{(B.5)}$$

where $\tilde{Z}$ is a normalisation constant. Note that the expression of the true posterior distribution of the hidden variables is the same up to replacing the true parameter variables with the corresponding tilde variables:

$$p(\mathbf{z}_{1:J_{tr}}, \mathbf{b}_{1:J_{tr}} | \mathbf{w}_{1:J_{tr}}, \mathbf{\Omega}) = \frac{1}{Z} \omega_{b_1} \left[\prod_{j=2}^{J_{tr}} \xi_{b_j b_{j-1}} \right] \prod_{j=1}^{J_{tr}} \prod_{i=1}^{N_j} \phi_{w_{ji} z_{ji}} \theta_{z_{ji} b_j} \tag{B.6}$$

Therefore, to compute the required expressions of the hidden variables $q(b_1 = b)$, $q(b_{j-1} = b, b_j = b')$, $q(z_{ji} = k, b_j = b)$ and $q(z_{ji} = k)$ one can use the same forward-backward procedure and update formulae as in the E-step of the EM-algorithm replacing all the parameter variables with the corresponding introduced tilde variables.

Appendix C
Distributions of Quadratic Forms

This appendix considers the distributions of quadratic forms of random vectors. These distributions are used in the proofs of the main theorems for the proposed statistical tests for change point detection given in Chap. 5. The proofs can be found in Appendix D.

Consider a distribution of a quadratic form $\mathbf{y}^T\mathbf{K}^{-1}\mathbf{y}$, where $\mathbf{y} \in \mathbb{R}^n$ is a random vector distributed as multivariate Gaussian and $\mathbf{K} \in \mathbb{R}^{n\times n}$ is a deterministic matrix. Two cases are particularly interesting in the context of the proposed statistical tests: the case when $\mathbf{K}$ is a covariance matrix of the random vector $\mathbf{y}$ and the case when $\mathbf{K}$ is an arbitrary symmetric matrix.

C.1 Quadratic form of the "Own" Covariance Matrix

Let

$$\mathbf{y} \sim \mathcal{N}(\mathbf{y}|\boldsymbol{\mu}, \mathbf{K}), \tag{C.1}$$

i.e., we are interested how the quadratic form of the covariance matrix of the given random vector is distributed.

Lemma 1 ([2], Chap. 14.6) *Let* $\mathbf{y} \sim \mathcal{N}(\mathbf{y}|\boldsymbol{\mu}, \mathbf{K})$. *Then*

$$\mathbf{y}^T\mathbf{K}^{-1}\mathbf{y} \sim \chi_n^{'2}(o), \tag{C.2}$$

where

$$o = \sum_{i=1}^{n}(\mu_i^{y'})^2, \quad \boldsymbol{\mu}^{y'} = \{\mu_i^{y'}\}_{i=1}^n = \mathbf{K}^{-\frac{1}{2}}\boldsymbol{\mu}, \tag{C.3}$$

and $\chi_n^{'2}(\cdot)$ *is a non-central chi-squared distribution with n degrees of freedom.*

O. Isupova, *Machine Learning Methods for Behaviour Analysis and Anomaly Detection in Video*, Springer Theses, https://doi.org/10.1007/978-3-319-75508-3

Proof Reorganise the input quadratic form:

$$\mathbf{y}^T\mathbf{K}^{-1}\mathbf{y} = \mathbf{y}^T\mathbf{K}^{-\frac{1}{2}}\underbrace{\mathbf{K}^{-\frac{1}{2}}\mathbf{y}}_{\mathbf{y}'} = \mathbf{y}'^T\mathbf{y}' = \sum_{i=1}^{n} y'^2_i, \tag{C.4}$$

where y'_i is the i-th component of the vector $\mathbf{y}'$.

Consider a distribution of $\mathbf{y}'$. Here $\mathbf{y}' \stackrel{\text{def}}{=} \mathbf{K}^{-\frac{1}{2}}\mathbf{y}$, where $\mathbf{y} \sim \mathcal{N}(\mathbf{y}|\boldsymbol{\mu}, \mathbf{K})$, therefore $\mathbf{y}'$ is also distributed as Gaussian. The following fact about the affine transformation of the multivariate Gaussian random vector is used: if

$$\boldsymbol{\zeta} \sim \mathcal{N}(\boldsymbol{\zeta}|\mathbf{v}, \boldsymbol{\Sigma}) \tag{C.5}$$

and

$$\boldsymbol{\zeta}' = \mathbf{A}\boldsymbol{\zeta} + \mathbf{p} \tag{C.6}$$

then

$$\boldsymbol{\zeta}' \sim \mathcal{N}(\boldsymbol{\zeta}'|\mathbf{A}\mathbf{v} + \mathbf{p}, \mathbf{A}\boldsymbol{\Sigma}\mathbf{A}^T). \tag{C.7}$$

In our case

$$\mathbf{y}' \sim \mathcal{N}\left(\mathbf{y}'|\mathbf{K}^{-1}\boldsymbol{\mu}, \mathbf{K}^{-\frac{1}{2}}\mathbf{K}\mathbf{K}^{-\frac{1}{2}}\right) = \mathcal{N}\left(\mathbf{y}'|\boldsymbol{\mu}^{y'}, \mathbf{I}\right), \tag{C.8}$$

where $\boldsymbol{\mu}^{y'} = \mathbf{K}^{-\frac{1}{2}}\boldsymbol{\mu}$.

As the covariance matrix is an identity matrix, (C.8) is equivalent to:

$$y'_i \sim \mathcal{N}\left(y'_i|\mu_i^{y'}, 1\right), \quad \forall i \in \{1, \ldots, n\}, \tag{C.9}$$

where $\mu_i^{y'}$ is the i-th component of the vector $\boldsymbol{\mu}^{y'}$.

From (C.4) and (C.9) the quadratic form $\mathbf{y}^T\mathbf{K}^{-1}\mathbf{y}$ can be represented as a sum of the squares of n independent normal random variables, which variance is equal to 1. According to the definition of a non-central chi-squared distribution, it gives:

$$\mathbf{y}^T\mathbf{K}^{-1}\mathbf{y} \sim \chi_n'^2(o), \quad o = \sum_{i=1}^{n}(\mu_i^{y'})^2. \tag{C.10}$$

□

C.2 Quadratic form of an Arbitrary Symmetric Matrix

Consider the case of a quadratic form of an arbitrary symmetric matrix, i.e.,

$$\mathbf{y} \sim \mathcal{N}(\mathbf{y}|\boldsymbol{\mu}, \mathbf{K}'), \tag{C.11}$$

and $\mathbf{K}$ is an arbitrary symmetric matrix.

Lemma 2 ([3]) *Let* $\mathbf{y} \sim \mathcal{N}(\mathbf{y}|\boldsymbol{\mu}, \mathbf{K}')$ *and* $\mathbf{K}$ *be an arbitrary symmetric matrix. Then* $\mathbf{y}^T\mathbf{K}^{-1}\mathbf{y}$ *has a generalised chi-squared distribution.*

Proof Transform the input quadratic form:

$$\begin{aligned}
&\mathbf{y}^T\mathbf{K}^{-1}\mathbf{y} = \mathbf{y}^T\mathbf{K}'^{-\frac{1}{2}}\mathbf{K}'^{\frac{1}{2}}\mathbf{K}^{-1}\mathbf{K}'^{\frac{1}{2}}\underbrace{\mathbf{K}'^{-\frac{1}{2}}\mathbf{y}}_{\mathbf{y}'} = \mathbf{y}'^T\mathbf{K}'^{\frac{1}{2}}\mathbf{K}^{-1}\mathbf{K}'^{\frac{1}{2}}\mathbf{y}' = \\
&\left(\mathbf{y}' - \mathbf{K}'^{-\frac{1}{2}}\boldsymbol{\mu} + \mathbf{K}'^{-\frac{1}{2}}\boldsymbol{\mu}\right)^T \mathbf{K}'^{\frac{1}{2}}\mathbf{K}^{-1}\mathbf{K}'^{\frac{1}{2}} \left(\underbrace{\mathbf{y}' - \mathbf{K}'^{-\frac{1}{2}}\boldsymbol{\mu}}_{\pi} + \mathbf{K}'^{-\frac{1}{2}}\boldsymbol{\mu}\right) = \\
&\left(\pi + \mathbf{K}'^{-\frac{1}{2}}\boldsymbol{\mu}\right)^T \mathbf{K}'^{\frac{1}{2}}\mathbf{K}^{-1}\mathbf{K}'^{\frac{1}{2}} \left(\pi + \mathbf{K}'^{-\frac{1}{2}}\boldsymbol{\mu}\right).
\end{aligned} \tag{C.12}$$

The matrix $\mathbf{K}'^{\frac{1}{2}}\mathbf{K}^{-1}\mathbf{K}'^{\frac{1}{2}}$ is symmetric as both matrices $\mathbf{K}$ and $\mathbf{K}'$ are symmetric. Therefore, the matrix $\mathbf{K}'^{\frac{1}{2}}\mathbf{K}^{-1}\mathbf{K}'^{\frac{1}{2}}$ is normal and the spectral theorem is valid for it:

$$\mathbf{K}'^{\frac{1}{2}}\mathbf{K}^{-1}\mathbf{K}'^{\frac{1}{2}} = \mathbf{P}\mathbf{D}\mathbf{P}^T, \tag{C.13}$$

where $\mathbf{P}$ is an orthogonal matrix and $\mathbf{D}$ is a diagonal matrix. Diagonal elements of $\mathbf{D}$ are eigenvalues of $\mathbf{K}'^{\frac{1}{2}}\mathbf{K}^{-1}\mathbf{K}'^{\frac{1}{2}}$ and columns of $\mathbf{P}$ are the corresponding eigenvectors.

Substituting (C.13) into (C.12) the quadratic form is further transformed as:

$$\begin{aligned}
&\left(\pi + \mathbf{K}'^{-\frac{1}{2}}\boldsymbol{\mu}\right)^T \mathbf{K}'^{\frac{1}{2}}\mathbf{K}^{-1}\mathbf{K}'^{\frac{1}{2}} \left(\pi + \mathbf{K}'^{-\frac{1}{2}}\boldsymbol{\mu}\right) \overset{\text{eq. (C.13)}}{=} \\
&\left(\pi + \mathbf{K}'^{-\frac{1}{2}}\boldsymbol{\mu}\right)^T \mathbf{P}\mathbf{D}\mathbf{P}^T \left(\pi + \mathbf{K}'^{-\frac{1}{2}}\boldsymbol{\mu}\right) = \\
&\left(\mathbf{P}^T\pi + \mathbf{P}^T\mathbf{K}'^{-\frac{1}{2}}\boldsymbol{\mu}\right)^T \mathbf{D} \left(\underbrace{\mathbf{P}^T\pi}_{\pi'} + \underbrace{\mathbf{P}^T\mathbf{K}'^{-\frac{1}{2}}\boldsymbol{\mu}}_{\boldsymbol{\mu}'}\right) = \\
&\left(\pi' + \boldsymbol{\mu}'\right)^T \mathbf{D} \left(\pi' + \boldsymbol{\mu}'\right).
\end{aligned} \tag{C.14}$$

Since $\mathbf{D}$ is a diagonal matrix, (C.14) can be expressed as:

$$\left(\boldsymbol{\pi}' + \boldsymbol{\mu}'\right)^T \mathbf{D} \left(\boldsymbol{\pi}' + \boldsymbol{\mu}'\right) = \sum_{i=1}^{n} d_{ii} \Bigg(\underbrace{\pi'_i + \mu'_i}_{u_i}\Bigg)^2 = \sum_{i=1}^{n} d_{ii} \underbrace{u_i^2}_{u'_i} = \sum_{i=1}^{n} d_{ii} u'_i, \tag{C.15}$$

where d_{ii} is the i-th diagonal element of the matrix $\mathbf{D}$, π'_i and μ'_i are the i-th elements of the vectors $\boldsymbol{\pi}'$ and $\boldsymbol{\mu}'$, respectively, $i \in \{1, \ldots, n\}$.

Combining (C.12), (C.14) and (C.15) the input quadratic form can be expressed as:

$$\mathbf{y}^T \mathbf{K}^{-1} \mathbf{y} = \sum_{i=1}^{n} d_{ii} u'_i. \tag{C.16}$$

Consider now distributions of all the introduced random variables. The same fact (C.5)–(C.7) about affine transformation of the Gaussian random variables is used as in the proof of Lemma 1. For the introduced random variables it is:

$$\mathbf{y} \sim \mathcal{N}(\boldsymbol{\mu}, \mathbf{K}') \tag{C.17}$$

$$\mathbf{y}' = \mathbf{K}'^{-\frac{1}{2}} \mathbf{y} \qquad \Rightarrow \mathbf{y}' \sim \mathcal{N}(\mathbf{y}' | \mathbf{K}'^{-\frac{1}{2}} \boldsymbol{\mu}, \mathbf{I}), \tag{C.18}$$

$$\boldsymbol{\pi} = \mathbf{y}' - \mathbf{K}'^{-\frac{1}{2}} \boldsymbol{\mu} \qquad \Rightarrow \boldsymbol{\pi} \sim \mathcal{N}(\boldsymbol{\pi} | \mathbf{0}, \mathbf{I}), \tag{C.19}$$

$$\boldsymbol{\pi}' = \mathbf{P}^T \boldsymbol{\pi} \qquad \Rightarrow \boldsymbol{\pi}' \sim \mathcal{N}(\boldsymbol{\pi}' | \mathbf{0}, \mathbf{P}^T \mathbf{I} \mathbf{P}) = \mathcal{N}(\boldsymbol{\pi}' | \mathbf{0}, \mathbf{I}), \tag{C.20}$$

$$\Rightarrow \pi'_i \sim \mathcal{N}(\pi'_i | 0, 1), \tag{C.21}$$

$$u_i = \pi'_i + \mu'_i \qquad \Rightarrow u_i \sim \mathcal{N}(u_i | \mu'_i, 1), \tag{C.22}$$

$$u'_i = u_i^2 \qquad \Rightarrow u'_i \sim \chi_1'^2(o_i), \quad \text{where } o_i = \mu_i'^2 \tag{C.23}$$

Therefore, from (C.16) and (C.23) $\mathbf{y}^T \mathbf{K}^{-1} \mathbf{y}$ is linear combination of independent chi-squared distributed variables. According to Definition 5.3, $\mathbf{y}^T \mathbf{K}^{-1} \mathbf{y}$ has a generalised chi-squared distribution. □

The quadratic form distributions presented in this Appendix are used in the proofs of the theorems for the proposed change point detections test (Chap. 5), which can be found in Appendix D.

Appendix D
Proofs of the Theorems for the Proposed Test Statistic

This appendix presents proofs for the theorems from Chap. 5.

D.1 Proof of Theorem 5.1

Proof Recall from (5.15) the test statistic can be expressed as:

$$T_{\text{online}}(\mathbf{y}_{\tau-L+1:\tau}) = (\mathbf{y}_{\tau-L+1:\tau} - \boldsymbol{\mu}_0)^T \left(\mathbf{K}_0 + \sigma_0^2 \mathbf{I}\right)^{-1} (\mathbf{y}_{\tau-L+1:\tau} - \boldsymbol{\mu}_0) + \log\det\left(\mathbf{K}_0 + \sigma_0^2\mathbf{I}\right) + L\log 2\pi. \tag{D.1}$$

Here $\mathbf{y}_{\tau-L+1:\tau} - \boldsymbol{\mu}_0 \sim \mathcal{N}(\mathbf{y}_{\tau-L+1:\tau}|\mathbf{0}, \mathbf{K}_0 + \sigma_0^2\mathbf{I})$. Therefore, the statistic T_{online} $(\mathbf{y}_{\tau-L+1:\tau})$ represents a quadratic form of the corresponding covariance matrix and an additive deterministic displacement. According to Lemma 1 (Appendix C):

$$T_{\text{online}}(\mathbf{y}_{\tau-L+1:\tau}) - \log\det\left(\mathbf{K}_0 + \sigma_0^2\mathbf{I}\right) - L\log 2\pi \sim \chi_L^{'2}(0) = \chi_L^2 \tag{D.2}$$

where χ_L^2 is a chi-squared distribution with L degrees of freedom. □

D.2 Proof of Theorem 5.2

Proof Recall that according to the definition (5.18) a power of the test is a conditional probability to reject the null hypothesis given that the alternative is true. The hypothesis $\mathcal{H}_0$ is rejected when $T_{\text{online}}(\mathbf{y}_{\tau-L+1:\tau}) < e_1$ or $T_{\text{online}}(\mathbf{y}_{\tau-L+1:\tau}) > e_2$, where e_1 and e_2 are the corresponding quantiles, determined in Definition 5.2, and the hypothesis $\mathcal{H}_1$ is true when $\boldsymbol{\vartheta} = \boldsymbol{\vartheta}_1$, i.e., $\mathbf{y}_{\tau-L+1:\tau} \sim \mathcal{N}(\boldsymbol{\mu}_1, \mathbf{K}_1 + \sigma_1^2\mathbf{I})$. The power of the test is then:

O. Isupova, *Machine Learning Methods for Behaviour Analysis and Anomaly Detection in Video*, Springer Theses, https://doi.org/10.1007/978-3-319-75508-3

$$\begin{aligned}B(\boldsymbol{\vartheta}) =&\mathbb{P}(T_{\text{online}}(\mathbf{y}_{\tau-L+1:\tau}) < e_1 \wedge T_{\text{online}}(\mathbf{y}_{\tau-L+1:\tau}) > e_2)|\boldsymbol{\vartheta} = \boldsymbol{\vartheta}_1) = \\ &\mathbb{P}(T_{\text{online}}(\mathbf{y}_{\tau-L+1:\tau}) < e_1|\boldsymbol{\vartheta} = \boldsymbol{\vartheta}_1) + \mathbb{P}(T_{\text{online}}(\mathbf{y}_{\tau-L+1:\tau}) > e_2)|\boldsymbol{\vartheta} = \boldsymbol{\vartheta}_1) = \\ &\mathbb{P}(T_{\text{online}}(\mathbf{y}_{\tau-L+1:\tau}) < e_1|\boldsymbol{\vartheta} = \boldsymbol{\vartheta}_1) + 1 - \mathbb{P}(T_{\text{online}}(\mathbf{y}_{\tau-L+1:\tau}) < e_2|\boldsymbol{\vartheta} = \boldsymbol{\vartheta}_1) = \\ &1 + \mathcal{F}_\beta(e_1) - \mathcal{F}_\beta(e_2),\end{aligned} \tag{D.3}$$

where $\mathcal{F}_\beta$ is a cdf of a random variable β and:

$$\begin{aligned}\beta = &T_{\text{online}}(\mathbf{y}_{\tau-L+1:\tau}) = \\ &\left(\mathbf{y}_{\tau-L+1:\tau} - \boldsymbol{\mu}_0\right)^T (\mathbf{K}_0 + \sigma_0^2\mathbf{I})^{-1} \left(\mathbf{y}_{\tau-L+1:\tau} - \boldsymbol{\mu}_0\right) + \log\det\left(\mathbf{K}_0 + \sigma_0^2\mathbf{I}\right) + L\log 2\pi\end{aligned} \tag{D.4}$$

Here $\mathbf{y}_{\tau-L+1:\tau} - \boldsymbol{\mu}_0 \sim \mathcal{N}\left(\mathbf{y}_{\tau-L+1:\tau}|\boldsymbol{\mu}_1 - \boldsymbol{\mu}_0, \mathbf{K}_1 + \sigma_1^2\mathbf{I}\right)$.

According to Lemma 2 (Appendix C), the first term in (D.4) has a generalised chi-squared distribution. The other two terms in (D.4) represent a deterministic displacement. Following the derivation of the proof of Lemma 2 β can be represented as:

$$\beta = \sum_{i=1}^{L} d_i v_i + \log\det\left(\mathbf{K}_0 + \sigma_0^2\mathbf{I}\right) + L\log 2\pi, \tag{D.5}$$

where d_i are eigenvalues of the matrix $\mathbf{A}$:

$$\mathbf{A} = \left(\mathbf{K}_1 + \sigma_1^2\mathbf{I}\right)^{\frac{1}{2}} \left(\mathbf{K}_0 + \sigma_0^2\mathbf{I}\right)^{-1} \left(\mathbf{K}_1 + \sigma_1^2\mathbf{I}\right)^{\frac{1}{2}}, \tag{D.6}$$

and v_i are random variables: $v_i \sim \chi_1'^2(o_i^2)$ while o_i are components of a vector $\mathbf{o} = \mathbf{P}^T\left(\mathbf{K}_1 + \sigma_1^2\mathbf{I}\right)^{-\frac{1}{2}}(\boldsymbol{\mu}_1 - \boldsymbol{\mu}_0)$, where $\mathbf{P}$ is a matrix, which columns are eigenvectors of the matrix $\mathbf{A}$. □

Appendix E
Optimisation of Gaussian Process Covariance Function Hyperparameters

This appendix provides details about optimisation of Gaussian process covariance function hyperparameters that is used in the proposed statistical tests for change point detection presented in Chap. 5.

Let $\mathbf{y} \sim \mathcal{N}(\mathbf{y}|\mathbf{0}, \mathbf{K}_{\boldsymbol{\vartheta}} + \sigma_{\boldsymbol{\vartheta}}^2\mathbf{I})$ be a noised observation vector of a Gaussian process (a zero mean function is taken for simplicity, optimisation of hyperparameters of a GP mean function is straightforward).

Consider the optimisation of a GP covariance function hyperparameter vector $\boldsymbol{\vartheta}$ by marginal likelihood maximisation [4]:

$$\log p(\mathbf{y}|\tau, \boldsymbol{\vartheta}) = -\frac{1}{2}\mathbf{y}^T\mathbf{K}^{-1}\mathbf{y} - \frac{1}{2}\log\det(\mathbf{K}) - \frac{N}{2}\log 2\pi \longrightarrow \max_{\boldsymbol{\vartheta}}, \tag{E.1}$$

where $\mathbf{K} = \mathbf{K}_{\boldsymbol{\vartheta}} + \sigma_{\boldsymbol{\vartheta}}^2\mathbf{I}$.

The optimisation of (E.1) can be performed by a gradient based optimiser. Consider the partial derivatives of the marginal likelihood:

$$\frac{\partial}{\partial\vartheta_j}\log p(\mathbf{y}|\tau, \boldsymbol{\vartheta}) = -\frac{1}{2}\mathbf{y}^T\frac{\partial}{\partial\vartheta_j}\mathbf{K}^{-1}\mathbf{y} - \frac{1}{2}\frac{\partial}{\partial\vartheta_j}\log\det(\mathbf{K}), \tag{E.2}$$

where ϑ_j is the j-th component of the vector $\boldsymbol{\vartheta}$.

Use formulae of matrix derivatives [4]:

$$\frac{\partial}{\partial\vartheta}\mathbf{K}^{-1} = -\mathbf{K}^{-1}\frac{\partial\mathbf{K}}{\partial\vartheta}\mathbf{K}^{-1}, \tag{E.3}$$

$$\frac{\partial}{\partial\vartheta}\log\det(\mathbf{K}) = \mathrm{tr}\left(\mathbf{K}^{-1}\frac{\partial\mathbf{K}}{\partial\vartheta}\right), \tag{E.4}$$

where $\frac{\partial\mathbf{K}}{\partial\vartheta}$ is a matrix of element-wise derivatives.

O. Isupova, *Machine Learning Methods for Behaviour Analysis and Anomaly Detection in Video*, Springer Theses, https://doi.org/10.1007/978-3-319-75508-3

Substituting (E.3) and (E.4) into the partial derivatives expression for the marginal likelihood (E.2) we obtain:

$$\frac{\partial}{\partial \vartheta_j} \log p(\mathbf{y}|\tau, \boldsymbol{\vartheta}) = \frac{1}{2}\mathbf{y}^T \mathbf{K}^{-1} \frac{\partial \mathbf{K}}{\partial \vartheta_j} \underbrace{\mathbf{K}^{-1}\mathbf{y}}_{\mathbf{y}'} - \frac{1}{2}\mathrm{tr}\left(\mathbf{K}^{-1}\frac{\partial \mathbf{K}}{\partial \vartheta_j}\right) =$$

$$\frac{1}{2}\mathrm{tr}\left(\mathbf{y}'\mathbf{y}'^T \frac{\partial \mathbf{K}}{\partial \vartheta_j}\right) - \frac{1}{2}\mathrm{tr}\left(\mathbf{K}^{-1}\frac{\partial \mathbf{K}}{\partial \vartheta_j}\right) = \frac{1}{2}\mathrm{tr}\left(\left(\mathbf{y}'\mathbf{y}'^T - \mathbf{K}^{-1}\right)\frac{\partial \mathbf{K}}{\partial \vartheta_j}\right). \tag{E.5}$$

The formula (E.5) for the partial derivatives of the marginal likelihood can be used for a gradient descent optimisation to find an optimal value for the hyperparameter vector.

References

1. C.M. Bishop, *Pattern Recognition and Machine Learning*, Information Science and Statistics (Springer, New York, 2006)
2. R.W. Keener, Theoretical statistics: topics for a core course, in *Springer Texts in Statistics* (2010)
3. H. Ruben, Probability content of regions under spherical normal distributions, IV: the distribution of homogeneous and non-homogeneous quadratic functions of normal variables. Ann. Math. Stat. **33**(2), 542–570 (1962)
4. C.E. Rasmussen, C.K.I. Williams, *Gaussian Processes for Machine Learning* (The MIT Press, 2006)

MIX
Papier aus verantwortungsvollen Quellen
Paper from responsible sources
FSC® C105338

If you have any concerns about our products,
you can contact us on
ProductSafety@springernature.com

In case Publisher is established outside the EU,
the EU authorized representative is:
Springer Nature Customer Service Center GmbH
Europaplatz 3, 69115 Heidelberg, Germany

Printed by Libri Plureos GmbH
in Hamburg, Germany